T0340812

The World of
ALL CREATURES
GREAT & SMALL
Welcome to Skeldale House

The World of
ALL CREATURES
GREAT & SMALL
Welcome to Skeldale House

Written by
JAMES STEEN

Foreword by **JIM WIGHT**

Michael O'Mara Books Limited

This updated paperback edition first published in 2022
First published in Great Britain in 2021 by
Michael O'Mara Books Limited
9 Lion Yard
Tremadoc Road
London SW4 7NQ

A CIP catalogue record for this book is available from the British Library.
Papers used by Michael O'Mara Books Limited are natural, recyclable products made
from wood grown in sustainable forests. The manufacturing processes conform to the
environmental regulations of the country of origin.

ISBN: 978-1-78929-483-5 in paperback print format
ISBN: 978-1-78929-407-1 in ebook format

2 3 4 5 6 7 8 9 10
Typeset by Ana Bjezancevic and Claire Cater
Cover by Ana Bjezancevic and Natasha Le Coultre
Printed and bound by CPI Group (UK) Ltd, Croydon, CR0 4YY

www.mombooks.com

*For the entire production team, cast and crew
who, in all weather and all circumstances,
go above and beyond to make this show.*

Contents

Foreword

BY JIM WIGHT

AUTHOR OF *THE REAL JAMES HERRIOT*

Around the autumn of 1968, I was watching my father tapping away on his little typewriter. I knew that he had been writing in his spare time for many years, and this resulted in a small book in which he recounted his experiences as a country vet in Yorkshire. I was also aware that he had forwarded manuscripts to various publishing houses, only to be rejected several times.

I had joined my father's practice in Thirsk about a year previously; as a young man setting out on his career, I realized that he displayed two qualities essential for success in any chosen field: determination and self-belief. He would not give up, despite the disappointing rejections, and he genuinely believed that his little book was worthy of publication. As he told me many times, 'If you consider yourself to be no more than third-rate, then that's all you will be; always aim high and have confidence in yourself.'

In the spring of 1969, his hours of hard work and dedication were finally rewarded when his book was accepted by Michael Joseph, the publishers in London. I well remember his excitement when a cheque for two hundred pounds arrived, his advance on publication. 'Look at this, Jim!' he cried, waving the cheque. 'This must be the pinnacle of my career!' Never could he have foreseen the scale of his future success. He wrote more books under the pseudonym of James Herriot, and they sold in their tens of millions worldwide. This modest and unassuming man, one who could never really understand the reasons for his success, had become the world's most famous veterinary surgeon.

So popular did the books become that they spawned two feature films followed by two hugely popular TV series, both titled *All Creatures Great and Small*: the first shown in the seventies and eighties, and the most recent – which introduces James Herriot to a new generation – currently being aired on Channel 5.

Our veterinary profession has changed enormously since James Herriot's day. His books are history, describing a way of life for the vet, farmer and countryman that has all but disappeared. This is one of the reasons for their enduring success; they are timeless, appealing to all walks of life, young and old, great and small. James Herriot's books are about people, they are about fascinating characters so vividly described that they almost jump out of the pages. More importantly, he writes about others, not just himself.

My father's life was greatly enriched by the hosts of interesting characters who crossed his path. Today, farms are full of modern machinery; in his day they were full of men, and many of the characters who would be portrayed in his bestselling books were based upon his experiences with them.

The old Yorkshire farmers were straightforward and plain-spoken. One farmer told my father that he had read one of his books, and on being asked whether he had enjoyed it responded with the words, 'Why, it's all about nowt!' On another occasion, following some 'bad news' in the village when the local shopkeeper's wife ran away with another man, a grizzled elderly farmer added his thoughts on the matter with, 'Aye. Ah wish someone would tek ma bugger ...'

One of the most vivid characters in the books is Siegfried Farnon, a character based upon my father's partner Donald Sinclair. A true character simply because he did not think he was. Donald had certain rules to be followed by the young assistants. Although very outdated, there is still a grain of truth in them today. One was: 'Always do something. If you think the animal is going to get better and needs no treatment, give it an injection, vitamins ... anything. If the animal recovers and you have done nothing you get no credit.' Another was: 'Paint a black picture. If the animal dies, you said it was going to die anyway ... if it lives following your treatment, you are a hero.'

My father's enduring advice to all new assistants was: 'The important thing to remember is that it's not what you do, but the way that you do it.' This was dramatically illustrated one day in 1940 during his time as an assistant to a veterinary surgeon, one J. J. MacDowall of Sunderland. He had been called to a difficult calving case, and had performed his task swiftly and professionally, delivering two live calves in about twenty minutes. He received no words of thanks from the farmer, only gloomy stares. Upon his return to the practice, he expressed his disappointment to his boss.

'Jock' MacDowall was highly experienced, one thoroughly acquainted with the art, as well as the science, of veterinary practice. 'You did a fine, professional job,' he said, 'but just think about it for a moment. They'd been trying to calve that cow for several hours and you come along and do the job in a few minutes. You got those calves out a bit too quick. You made them look a bit silly. And they are paying us good money, so never make a job look easy.'

He finished with words that my father never forgot. 'Remember, it's not *what* you do, it's the *way* that you do it. I can tell you that on countless occasions when I have been delivering calves, the sweat has been pouring off me holding the bloody things in.'

The character of Donald Sinclair has been so very well portrayed on TV, by Robert Hardy in the BBC series and by Samuel West in the current show. Both men have effectively captured the qualities of that great character: his impatience, unpredictability, volatility, but above all his down-to-earth decency. Donald's eccentricities meant that he was a very difficult man to work with. I remember asking my father one day how he had managed to work with such a man for over fifty years. 'What laughs I have had with him!' he replied. 'And I have always known that he would never stab me in the back. There are no hidden agendas with Donald. And what a wonderful character to put in my books!'

As I watch the TV series on Channel 5, many memories come flooding back. The small family farms, most of which have now disappeared, and the hard-working farmers labouring all the hours that God sent; the vets visiting so many farms at all hours of the day and night. The Yorkshire farmers are well represented on the TV screen, both in their appearance and the way they act their parts.

These memories are all enriched by the superb acting. Nicholas Ralph plays a splendid part as James Herriot – a difficult task, as my father always considered himself a 'grey figure' among all the wonderful characters. Like Chris Timothy in the BBC series, he achieves just the right level of authority – centre stage but not in any way dominant. Rachel Shenton is extremely good as Helen. When I first met her, I thought that despite her very appealing nature and appearance, she would never pass as a farmer's daughter. Rachel has proved me wrong.

Callum Woodhouse portrays Tristan really well, and has a difficult job being compared to Peter Davison in that role for the BBC. Like Peter, though, he gets well into the character of Tristan – real name Brian Sinclair, and a man I knew well. The wayward and fun-loving younger brother of Siegfried is brought to life on the screen. Anna Madeley steps admirably into the role of Mrs Hall, the housekeeper, a minor character in the Herriot books, but one which has been transformed into a major player in this show.

The series is extremely professionally produced and directed. The whole ethos of veterinary practice in those days is well portrayed. The set has brought memories flooding back, as I remember my childhood in the old rambling number 23 Kirkgate, Thirsk – immortalized by Herriot as Skeldale House. That house in Thirsk is now The World of James Herriot, a museum and visitor centre. Thousands of Herriot fans from all over the world have been through its doors and I feel sure they will continue to do so.

This book gives a most informative and fascinating insight into the hard work, organization and dedication that goes into the production of a television series such as this. The author reveals the underlying problems that beset the various departments, the producer, the director, location manager and production designer, to name but a few. An additional hurdle to be faced is the involvement of animals, with animal-welfare experts and animal handlers needing to be involved. I am especially pleased that Andy Barrett – who began his career working with me in Thirsk – is enjoying his time on the set as the veterinary consultant.

The World of All Creatures Great and Small is extremely informative and very easy to read, and I recommend it not only to all James Herriot fans but also to anyone who enjoys watching a good television production. The description of the work involved behind the scenes makes for an engrossing read.

PART ONE
Beginnings

'Do you ever wonder what your life would've been
like if you hadn't've come to Darrowby?'

– HELEN ALDERSON TO JAMES HERRIOT

CHAPTER ONE

The Bond

Two moments in time. Two scenes. Both took place in Yorkshire on 17 September, yet they are separated by precisely forty-eight years. Without the first, there would not have been the second.

The first: 17 September 1971. Alf Wight – aged fifty-four and perhaps at his now-famous veterinary practice in Thirsk, close to the Dales – takes a pen and signs a contract. The document seals his deal with the New York publisher St Martin's Press for a book that will be entitled *All Creatures Great and Small*. The book is written under Alf's pen name, James Herriot. It chronicles the escapades and adventures of James, and comprises Herriot's debut book, *If Only They Could Talk*, and his second, *It Shouldn't Happen to a Vet* (both published in Britain by Michael Joseph).

They may read as novels but really the Herriot books are memoirs. They are accounts of Alf's real-life experiences, with characters often a composite of the people he had met, worked with and for. Had he written the book under his own name it would have been deemed advertising, and therefore breached the veterinary profession's rules and etiquette. But why the name James Herriot? One evening, while watching a football match on television, it came to Alf. This was the name of the Scottish goalkeeper who was there, on-screen, playing for Birmingham City against Manchester United. He didn't make it up.

On the face of it, Alf's story – witty and great fun, often hilarious, frequently poignant – is of a vet, in his twenties, who comes to the market town of Darrowby in Yorkshire. There he lives in Skeldale House, a veterinary practice and home, with the Farnon brothers, Siegfried and Tristan, as well as the rarely mentioned housekeeper, Mrs Hall. On the farms and in the villages, young James sets out to learn and master the skills of a vet. Somehow he must cope with the testy, unpredictable ways of his employer, Siegfried, based on Alf's employer, Donald Sinclair. And somehow he must endure the unrelenting chill and rain, and all the time try to forge and maintain relationships with his clients, the farmers who – Yorkshire-style – really do tell it like it is, calling a spade a spade.

Do not for a minute underestimate the harsh realities of this vet's life in Yorkshire. Alf's son, Jim, himself a vet (now retired), says: 'Dad was taking me out to farms when I was about three years old, and I remember his cars. No brakes, no heater. Having to drive up to the Dales in winters which were proper winters. I don't know how the hell he did it. An hour's drive in sub-zero temperatures and he'd arrive at the farms stupefied with cold. That was his word for the way he felt – *stupefied*. Before he could even function he'd have to go into the kitchen to thaw out. The farmers were all right. They'd already got layers of coats on, and had been mucking out and exercising and milking the cows.'

Look more closely at the Herriot stories and there is a profound depth, and one that engages anyone of any age. Very cleverly, he presents a highly observant study of human nature and human strength, struggle and frailty. Chapter after chapter, he takes the reader on a voyage over life's waves – the ups, the downs – of hope and despair, rejection, failure and success. You do not need to be a vet or indeed an animal lover to appreciate and enjoy his writing. These are tales to make you laugh out loud and shed a tear, and all based on Alf's life. 'The characters,' says Jim, 'bounce out of his books.' Herriot reassures us of the importance of bonds – between people, between people and animals, between people and land. Family, community, unity, friendship: the brushstrokes that form James Herriot's vivid picture of life in Yorkshire in the thirties.

Possibly wishing to make the most of the love angle, the publishers in New York had a request for Alf. Would it be possible, they wondered, for him to write the happiest of endings, one in which James Herriot marries Helen? Alf agreed; he liked the notion and the idea that the adventures of James Herriot were soon to be enjoyed in America. (The USA was then *the* place to be: a fortnight before Alf signed the contract, John Lennon had left Britain to live in New York.)

The sales of James Herriot's books are in the tens of millions. Despite his remarkable success as a writer, Alf remained a vet and lived in Yorkshire until his death in 1995, outlived by his beloved wife, Joan, his inspiration for the character of Helen Alderson. Jim says, 'When he'd made money as James Herriot – and he was not a greedy man – I said to him, "You were never interested in money, were you?" And Dad said, "I was once. When I didn't have any."'

Alf was always cheerful when approached by the many Herriot fans who came from all corners of the globe to meet the man who had entertained and inspired them, and shake his hand. They had read his books, over and over again. And how many of his young readers, you wonder, were inspired to become vets? Or maybe they had seen the 1975 film adaptation, or the BBC series of All Creatures, a staple Sunday night treat from the late seventies and into the eighties. That was the first TV adaptation. Then there was a second …

The second moment: again 17 September and now in the heart of the Dales, though the year is 2019. This National Park has not changed much since Alf's day. A film crew is busy at work, and a vintage Leyland Lion bus (built in 1927, when Alf was eleven years old), rolls along an otherwise empty road, deep in the countryside. There are two young actors on board, Nicholas Ralph and Rachel Shenton. They are filming the first episode of *All Creatures Great and Small*, the second adaptation for television. And they are shooting the particular scene in which James meets Helen for the first time, a scene of which Alf and his US publishers would have thoroughly approved.

There they were, gathered as a unit, created to make a six-part series and a Christmas special. (One year later, in September 2020, that first episode would be broadcast for British viewers on Channel 5 and, a few months later, in the States, on Masterpiece on PBS.) Around them, the rugged beauty of the Dales. Remove from this picture the two man-made features – that road and the vintage bus – and this expanse of landscape could have been just as it was in the Iron Age. The road leads to the nearby village of Malham where, many centuries ago, the stream was a magnet, attracting people because they regarded water almost as a God, giving life and longevity to man and beast.

This led to a small settlement, listed as Malgun in *The Domesday Book*, commissioned by William the Conqueror and completed in 1086.

Not too far from cast and crew there were the great limestone landmarks of the Dales that are nature's sculptures: Malham Gorge, a massive cliff formation in the shape of an amphitheatre; and Gordale Scar, a gorge etched in cliffs, with waterfalls that cascade from a height of several hundred feet. The gorge is said to have inspired Helm's Deep in J.R.R. Tolkien's *The Lord of the Rings*. This is a part of the world that has long attracted walkers, ramblers, cyclists and hikers … and film-makers; scenes were filmed here for *Harry Potter and the Deathly Hallows: Part One*.

A couple of months earlier, Brian Percival, the lead director, had driven along the same stretch of what is known as Malham Road with Gary Barnes, the location manager. As they took in the view of Malham Lings, a natural formation of limestone cracked 'pavements', they agreed that somehow this setting should – *must* – feature in an episode of All Creatures. It was simply too stunning to miss. Now, here they were with the cast and crew, cameras and lights.

The same road, which weaves through the dips and peaks, appears to come from nowhere and go nowhere, and it also features in the scene in which James steps off the bus after believing it is the stop for Darrowby. As the camera pans around, we see that he is a solitary figure in the middle of the Dales. He is isolated at a crossroads. 'We put him at a crossroads because it's a metaphor for his life,' says Brian. 'It was only later that I was watching *North by Northwest* and was reminded of that scene – a classic – of Cary Grant standing at a crossroads. I thought, "Oh God …"' And this crossroads motif is repeated in the Christmas special, with James in the car on his way to Glasgow, before turning the car around and returning to Darrowby for Helen's wedding.

Before filming that day, at 6.30am, Gary Barnes stood and watched as the darkness retreated. The sun rose over the hills and lit up and began to warm the Dales. He said to himself, 'If I've got this view for the next few months then I'm a very lucky man.'

This day was just as memorable for Erik Molberg Hansen, the Danish-born director of photography (DP). Erik recalls, 'We had been to that location two or three times before, driving out in a couple of buses to do technical recces. It was seriously windy and one time it was raining so hard we said, "Quick!" and we ran for the cover of our buses. We planned these scenes very well, going through it and through it again. Then the day came to shoot, and the sun came out. The Yorkshire Gods gave us good luck.'

Knowing that Helen needed to stand out in the landscape, costume designer Ros Little had dressed her in a red coat. 'It's such a special scene,' says Erik. 'The green, the red coat, the red bus, and interaction between people and landscape.' The Yorkshire Gods had another gift. As James began his walk to Darrowby, a gust of wind suddenly blew his hat from his head. Perfect! 'It was just by chance,' says Erik. 'And in the first take! We were so lucky.' Irony of ironies, the weather was so good that a rain machine was required to create a scripted downpour as James began his walk to Darrowby. That's showbiz!

CHAPTER TWO

The Focus

Seventy-nine years after Alf Wight had left his home in Glasgow and travelled to Yorkshire to take up life as a veterinary surgeon, Nicholas Ralph left his home, also in Glasgow, and made his way to the Dales to play the part of Alf's alter ego. (One of the early scenes in episode one would show Nick as James, at Glasgow railway station and boarding a train for the Dales.)

The shoot began on Saturday 14 September. However, in the preceding days four of the main cast were meeting in Skipton, known as the southern gateway to the Dales. They were: Nicholas Ralph, a newcomer to the screen and playing James Herriot; Samuel West (Siegfried Farnon); Callum Woodhouse (Tristan Farnon); and Rachel Shenton (Helen Alderson). Over a few days, they would rehearse with a read-through of the script. Anna Madeley (Mrs Hall), joined the cast later, in the first week of filming. She was accompanied by her baby.

Nick's journey required taking a train from Carlisle to Skipton, a two-hour journey on a track that goes straight through the Dales. 'I'd never been to the Yorkshire Dales,' Nick recalls, 'but in preparation for the role, I'd read *All Creatures Great and Small*, in which the area is described beautifully.' He is from the north of Scotland, and 'it's beautiful up there'. So he reckoned that while the Dales would be lovely, it couldn't be much better than the Highlands where he grew up. Peering through the window of the train carriage, he was actually taken aback by the landscape. 'This is insane,' he thought. It was like a painting – the rolling Dales scattered with sheep and cattle, hemmed with drystone walls, and dotted with the occasional solitary silhouette of a tree on the peak of a Dale.

For Nick, this was more than a train journey. He was on his way to becoming an accomplished television actor. It had begun, some months earlier, in the spring when he auditioned for the part of the main character, the young vet. Beverley Keogh and David Martin, the show's casting directors, had gone to Scotland to find a James Herriot for the series. In the book James is Scottish, and Alf Wight had a Scottish burr all his life. A genuine Scottish actor was necessary, says Beverley, and not simply a Scottish actor but one that 'embodied the complete characteristics of James Herriot himself – a terrific sense of fun, a twinkle in his eye, a strong work ethic and an appealing humility'.

Nick had been among scores of new actors who had been auditioned. With Nick, however, they had 'an inkling that he could bring something really interesting to the piece. He had not worked in television or film, but he had been working hard at his craft. He was an alumnus of the Royal Conservatoire of Scotland and gave an outstanding performance in the play *Interference*.'

Beverley adds, 'Auditioning for such a well-known role can be a daunting process for any young artist. Sometimes nerves come into play, and you have to take that into account and work with them to allow for a more comfortable performance.' But with Nick there was none of that. 'He was effortlessly natural and instantly likeable.'

Once shortlisted, he had been invited to 'self-tape' with additional notes from the All Creatures' team, and then attend a meeting to be put through his paces by Brian Percival. Working with Brian, Nick continued to elevate his performances. Beverley recalls, 'His take on James grew more layered and nuanced each time he was brought back to audition, until there was no arguing that Nick was everything we were searching for, and more. A true breath of fresh air.'

Meanwhile, Samuel West, Callum Woodhouse and Rachel Shenton were making their way from London to Leeds railway station, the busiest station in the north. Carrying enough luggage to see them through a few days – and, in Rachel's case, a yoga mat – the trio had met by chance. They boarded the local train that took them north, out of the city and into the Dales and Skipton. If it is true that it is the journey that counts not the destination, then this forty-five-minute train ride prompted a bonding and friendship. 'We had,' says Sam, 'the most delightful chat.'

'It's funny,' adds Rachel, 'because you just have a list of names at that point, and of course I knew about them professionally. I was really excited, but we were strangers, didn't know each other at all. I collared Callum on the train and said, "Oh hi, you're playing Tristan. I'm playing Helen. It's lovely to meet you." Then we chatted about dogs, which was very nice and is still what we talk about all the time.'

This became the happiest of film sets, and that sense of camaraderie was evident from the start. After that initial read-through Nick struck up a friendship with Callum, who made his name in the ITV adaptation of *The Durrells* (like All Creatures, the series is set in the thirties, though beneath the shimmering sun of Corfu).

On that day when they met in Yorkshire, the two young actors had their read-through and then, come the evening, went for a beer in their hotel. The minute they met, they clicked. Nick was keen to learn; Callum happy to share his knowledge. 'I think I asked him every question under the sun about acting for the screen,' says Nick. 'He was very generous and I'm sure I bored him to death. But we had a great laugh. He really put me at ease.' Callum adds, 'I was nervous and excited about All Creatures, but with Nick asking all these questions, it sort of helped alleviate some of the nerves.' He also told him, 'Look, you've been to drama school and you've done the auditions. You've got the role. You're obviously good enough so it'll all be fine.' These words gave Nick a confidence boost. And of course, as it transpired, Callum was right.

In fact everyone was keen to help Nick, and he was eager to absorb advice from his co-stars and the crew. Deeply curious and inquisitive, he had a seemingly endless list of questions for the cast. Sam put him at ease on the first day. 'Sam is a bird lover,' Nick explains, 'and I'd happened to say to him that, when I was younger, I liked birds, particularly chaffinches. So there we were, both in the car, and I was feeling focused. "Rolling! Turning! Sound! Speed!" And then suddenly Sam shouts out, "Chaffinch! I just saw it pass, so I had to tell you."' Who knows if he had indeed seen it, but he certainly put a smile on Nick's face.

On the first day of filming, Sam also took a photograph and posted it on Twitter. It showed the clapperboard, adorned with a picture of a cow and two sheep, with the words: 'Scene one (21, 22, 23), slate one, take one. Director: Brian Percival. DP: Erik M Hansen.'

Soon Nick would be 'up close and personal with farmyard animals'. The young actor almost had to pinch himself. It was not long before he and Rachel were working with a calf that had a (fictional) suspected broken leg. 'So you've got Rachel there, kneeling down with the calf's head on her lap, stroking it and making sure it's calm and all right. Meanwhile, I'm going through the procedure of establishing whether the calf's leg is damaged, with its mother looking over. I'm thinking, 'Here we are, in a barn in the middle of the Yorkshire Dales. Surreal.' Rachel remembers: 'As I was stroking the calf it was going to sleep – it was meant to be in pain!'

Nick was well into his stride early on in filming, which would have come as no surprise whatsoever if you had seen him steal the show at the Nairn Highland Games in the mid-nineties. 'I was about four years old, there was a band playing, and I sneaked through a fence. I started doing what I thought was Highland dancing. I was a show-off and people took pictures of me, which ended up in the paper.' Four years later he auditioned to appear as a breakdancer on a TV talent show, 'so I've always had a thing for performing'.

Brian Percival knew straight away that Nick was the perfect James. 'Yes, he was young and inexperienced in that he had never been in front of a camera before. And so yes, I gave him advice. I always try to be helpful rather than say, "Do this, do that." But we just hit it off from the start. And he wanted to learn from me. I wanted to embrace his natural talent. We had found someone incredibly special and unbelievably likeable and modest, one of the nicest guys you'd want to meet, and so willing to learn and do whatever he could. He was on it 110 per cent.'

'I don't think I have come across anyone in my career who is so natural in terms of working with a camera. He knew where his light was, where he should be. All those things sometimes you have to give an actor. I love working with young actors. One of the most satisfying things you can do. Quite often you see the change and it really makes the job worthwhile. You feel you have given something worthwhile to someone, and in return they give it back to the audience.'

The night before filming began, Nick's nerves were jangling, and he'd been terrified that he would oversleep. 'I set about ten alarms and then you wake up on the hour, every hour. A great sleep, eh? But it didn't matter because I was buzzing, full of energy. And the next morning we got dropped off at the unit, with all the trailers and the catering trucks. I got offered one of the exec's trailers, and I was like, this is so cool.'

And Sam? 'On day one we were all beautifully nervous – I mean nervous in the right way.' And Rachel admits, 'I hadn't slept for two nights before we started filming.'

Brian Percival, even with his wealth of film-making, also had an bad night's sleep. 'There's always nervous energy,' he says. 'Always a sleepless night before we start a shoot … and yet that feeling, knowing that everything is done and under control.'

A few days before filming, and in the Dales, he had stuck to his recipe – tried and tested umpteen times – for a good start. 'I like to get everything tied up and in place. Everybody rehearsed, and we know where the lights are going, and everybody knows exactly what they are doing so that when we get to the point of filming we are ready. No panic. I have everything planned.'

His aim always is 'to almost relax into the filming … And then my attitude is that we'll do what we've planned unless we find something better. Someone – it doesn't matter who – might come up with an idea as we're filming. So I don't care where the idea comes from, and if it's better than the plan then we can use it. But if it isn't better, at least we already know that we are going to get something that's good and is going to work.' This means that on the first day of shooting, 'No one is standing around saying, "What would Nick's hair look like with less Brylcreem?"'

Brian commanded the respect of the cast and crew. Professionally, the Liverpudlian is well established with vast experience. He was the lead director on ITV's *Downton Abbey*, and instrumental in its success. (As with All Creatures, Downton takes the viewer back in time and revels in relationships.) Brian is definitely not the caricatured director, the one who booms and barks orders at a fearful cast and crew.

Those who have worked with Brian – they include Anna Madeley – speak of him with immense affection. He is 'gentlemanly', 'kind' and 'softly-spoken'. 'As with so much in life, patience brings rewards,' he says. 'If you panic, it's a waste of energy. If you scream and shout, it isn't going to get you anywhere. So you just have to be very calm and very patient and think about relaxing the cast. That way, people will go into filming feeling good about it. If your first day is hell they'll think, my God, we've got sixteen weeks of this.'

CHAPTER THREE

Challenges

There'd be challenges ahead. And not so far ahead, either. A few days into filming they were shooting the scene in episode one in which James, who has come for his job interview, is taken to the farm by Siegfried. The young vet has to wade through the mud to inspect an abscess on a horse's hoof. Siegfried is putting him to the test.

'We filmed until eleven in the morning and it was all going well,' recalls Brian. 'And then the heavens opened up. The rain did not stop for about two hours, so heavy that we couldn't stand outside. Of course everything that had been shot before the downpour showed dry walls, dry houses, dry landscape. That was in the first week and so, right at the end of the shoot, we had to go back to that same farm and spend half a day completing the original scene when it was dry again.'

The conditions can change from sun-baked to sodden at shocking speed. However, as location manager Gary Barnes says, 'We could have filmed more in the studio, but it was crucial to show off Yorkshire. So we had to learn to live with the weather. Simple as that.'

Those heavens opened up throughout most of the filming of the Darrowby Fair in Grassington. A good deal of planning had gone in to creating some spaces under cover, such as the village hall and filming the pet show in Skeldale House. The constant downpour in Grassington meant the crew were wet and cold, but the visuals didn't suffer, thanks to Jackie Smith's joyous, colourful design and the skills of director Metin Hüseyin and Erik Persson, director of photography. Ben Vanstone says, 'It was horrendous weather when we filmed the fair, but it still looked fantastic. For me, that was the moment when it felt we were really doing something special.' Even in the rain, Yorkshire is beautiful and uplifting. Contrast this incessant, pelting rain with the final few weeks of filming the second series, when the sun shone gloriously over the Dales and cast and crew felt blessed.

Those early days brought another headache. When the shoot began the studio set was still being built, thanks – strangely – to Brexit. The set for Skeldale House, they had decided, would be created within a warehouse. However, as Britain prepared to leave the EU businesses feared shortages, and were buying up warehouses to stockpile their goods. Shortly before filming began the producers had indeed found a warehouse – near Harrogate – but it was a last-minute find and the set was receiving the finishing touches. 'It was a luxury we lacked at the beginning,' says Brian.

Melissa Gallant, the executive producer, says, 'I do think that our biggest challenge of all was making a new adaptation of the books following in the footsteps of a very long-running, hugely beloved show that was still fresh in the minds of the nation. It was something of a national treasure, a show that came with the happiness and warmth of nostalgia – a family favourite that evoked memories of being allowed to stay up in your pyjamas to watch it. For a lot of people it was perfect, and the idea of a new version really didn't appeal.'

When Channel 5 announced that it had commissioned this new adaptation there were many worried Herriot fans. 'I think people genuinely feared it wouldn't be as good, and didn't feel that a new series was necessary. We understood. Seeing your favourite book adapted, or your favourite film remade and hating it is a truly disheartening experience. A lot of people didn't want a new version because they still loved the previous one, and didn't feel that it needed to be re-made. We had an awful lot to live up to – for the audience, for the Herriot family and for Yorkshire.' Would an audience accept another *All Creatures Great and Small*?

The producers were keen to bring the Herriot stories and characters to a new generation who probably hadn't read the books and, at the same time, make a show for those who already knew and loved All Creatures. Should they try to replicate elements of the BBC series, or do something completely fresh? It was easy to update the visual world of the show with modern cinematography and drones, but what about casting? 'Our job,' says Melissa, 'was to replicate the feeling that you get when you read the books, the same feeling that the BBC series evoked. We had to deliver the same experience. So … nostalgic but new, with a contemporary approach to storytelling. And we wanted to explore Herriot's characters in greater depth, in a way that would sit respectfully alongside our predecessors. It had to be confidently new without turning its back on the things people loved about All Creatures. If the audience didn't come back for the world and the characters then we'd have got it wrong.'

Looking back now, she is in a position to say, 'When it went out one of the genuinely biggest delights was that so many people who'd loved the BBC series now accepted and appreciated this new one. Our cast had become those characters within seconds of being on the screen.'

'Never work with animals or children,' said W. C. Fields, that great comic actor of the movies' Golden Age. (Fields, an unstable, changeable, heavy drinker, could have added his own name to the list of no-nos.) So here they were, seemingly oblivious to Fields' well-coined axiom, embarking on a series that would feature plenty of children and even more animals. The title is the giveaway. And what happened behind the scenes?

Sam describes the call sheet on a television or film job as 'a sort of magical spell'. 'It gives you the names and the telephone numbers of sometimes a hundred people who do extraordinarily varied jobs, all to the highest standard. It gives the time when they are meant to turn up, and the place where they are meant to be. And amazingly they do turn up exactly when they are meant to, and they do the things they are brilliant at. And the rest of the time they melt away, and are completely invisible and totally inaudible and utterly concentrated.

'And at the end of the day you might just have canned something of lasting worth. You might have only done forty-five seconds of slightly boring driving shots. But you probably would also have done something that made you think, yeah that *was* a good day.' (He adds that, with All Creatures, there was always the possibility of canning something of lasting worth, 'and that's a great feeling to go to work with'.)

The call sheet for that first day of filming on All Creatures was unusual – yes, even perhaps for veterans like Sam who think they have seen it all. The document included the requirements for furry cast members: 'Misty the cat, Leo the cat, Jasper the cat, a rabbit, a tortoise and one small dog.' The tone was certainly set for what was to come. On that first day there was an animal handler, animal welfare expert, on-set vet and a tutor for the children.

Brian remembers a moment in 2003, just before filming began on *Pleasureland*, his film for Channel 4. 'The cast was a bunch of kids who had never acted before. They were about seventeen years old. The night before filming, Katie, the star of the film, said to me, "You know you're doing this film with all of us who have never acted before ..." I said, "Yeah." "Don't you think you're taking a bit of a risk?" And I thought, "Well, now you mention it. Thank you, Katie!"' He adds, 'The thing is, you get so involved in the story. It's only later you might think, "Oh dear."'

'Working with animals was always going to be one of the more interesting challenges,' is how producer Richard Burrell puts it, with his characteristic understatement and the air of someone who is rarely fazed. 'Small creatures perhaps were all fairly familiar, but when we're dealing with sheep and horses and cattle that's a very different area.'

Film-makers have a responsibility (legal as well as ethical) to consider animal welfare. There are rules and regulations. 'In story terms, the animals are unwell, sick or distressed,' Richard says. 'So we must create that for the camera and the audience, but make sure the animals we use are never distressed. We do it in a way that is responsible and fully compliant with the laws, and morally how one would want to do it.'

To achieve this, they created 'a triangle' of expertise and skills that covered three key areas: first, using those who provide and train the animals; second, a consultant vet to be on-set to advise the actors and crew on procedures, and to be a double when an actor was not permitted to carry out a clinical procedure; and third, an animal welfare expert, Jody Gordon, again on-set. 'We'd have meetings when the scripts come in and we'd say, "Right, we need this cow that's going to lie down, or this horse which is supposed to be distressed."'

The animal handlers are 1st Choice Animals, run by the mother and son team, Jill and Dean Clark. 'Essentially they made the animals better actors than us five,' says Callum, referring to himself, Nick, Sam, Anna and Rachel.

The first consultant vet was Jim Wight, the son of Alf. Jim is a retired vet, who worked with his father at the practice in Thirsk, at 23 Kirkgate on which Skeldale House is based. He continues to read the scripts, spot potential howlers and offer advice. When filming began, he was asked if could recommend a vet who could be on-set. Jim suggested Andy Barrett, who had also worked with Alf Wight in the eighties, and lived briefly in a room above the practice. Andy not only gives advice about veterinary procedures before filming and advises the actors (helping them to look convincing as vets during procedures), but is also a stand-in for Nick, adeptly performing procedures for the camera – although it is only Andy's hands and forearms that you see in shot.

Before filming started, Andy had a 'boot camp' with the main cast. He took them to a number of farms in the Dales, and they got up close with the animals and asked: How do you approach a horse? How do you lift a horse's leg? How do you take the temperature of a cow? Sam would later say, 'The thing I learned quite quickly was to get out of the way of anything that was standing up, whether a cow or a horse or even a calf that wanted to move. I've got two young children and I'm used to being trod on. But a calf is not a five-year-old. It weighs the same as a small car and, yeah, I was pleased with my hob-nailed boots. They saved my toes on more than one occasion.'

The animal welfare expert, Jody, is also involved at the planning stage, answering the producers' questions. What is possible? What is not possible? What can we do? What shouldn't we do? Animals, just like humans, need breaks. These would be built into the filming schedule.

Sharon Moran, production executive at Playground (the TV and media company), already had a significant, vital role in the production. She also took on the mantle of supervising the 'triangle', overseeing all things animal on the show along with the handlers and advisors (both veterinary and welfare). The regulations concerning the movement of animals presented a learning curve up there with the Yorkshire Three Peaks (walking) Challenge.

The movement rules are designed to prevent the spread of disease and, weirdly perhaps, are similar to the Covid regulations. Sharon explains, 'There are standstill regulations which stipulate that once you have moved cattle, sheep or goats to a farm they have to remain there for six days. A pig has to stay there for twenty-one days. If an animal is poorly it is likely to have to stay there longer and will receive visits from the vet at the standstill location.' During this standstill period, the animal handlers visit the animals at least twice a day to check, feed, water and clean out their living quarters.

Yet there is always the potential for the knock-on effect when filming. 'Shooting schedules are complex and take an immense amount of planning as there are so many elements to take into account,' Sharon adds. 'We're juggling the availability of actors and locations, the weather and light, and when animals are on standstill, that affects everything. Scenes are shot around the availability of the animals and their schedule.'

Erik Molberg Hansen, the director of photography, says animals are like the weather. 'Both of them can make film-makers very stressed. But you mustn't get angry. Stay relaxed and remember, the animals didn't ask to be there. They're not getting paid. So you try to make sure everything is well prepared. It's like filming with young children. You have the lights ready and so on. You don't work your way towards it. There is plan A. Then there are plans B, C, D and E. And at the end of the day you might say, "Maybe we didn't get everything we wanted but perhaps we can find another half day to get a bit more."'

Richard and Brian also asked themselves – understandably – if there were times when they could avoid using animals in the shot. Experts in prosthetics are so skilled that they can make almost anything fake look real. Pauline Fowler was appointed. Her job is to make and mould those bits that seem to be genuine animal parts but are, in fact, made from fibreglass or silicone. Her appointment began with the question: 'Can you make a cow's neck?' Later, in series two, the requests included: 'Can you make a boil for the ear of Buttercup the pig.' Oh, the magic of prosthetics! As viewers, we believe we see an animal or part of it. In fact the animal or part of it was not filmed, but extremely well faked.

Consider, for a moment, the birth of the Border collie puppies in the Christmas special of All Creatures, directed by Andy Hay. As viewers, you and I see … well, the birth. But behind the scenes this screen magic was created with two dogs, quite a lot of silicone and a team of experts that included an animal handler, an animal welfare specialist and a prosthetics specialist. Oh, and a carpenter to make two whelping boxes; a safe space for the puppies to be born in.

Betsy was the pregnant mother. Jazz was the stand-in mother and not pregnant, and was used as a double for the action scenes with the actors. Faced with this challenge of achieving a live puppy birth on camera, the animal, design and production teams set to work, collaborating on how this would be done while preserving the welfare of the animals.

Many weeks before the pups' delivery, a designated puppy delivery suite room was set up on the production's farm. A green screen was placed on a wall of the suite, along with a custom-made whelping box. The expectant dog uses the box as a nest, and then gives birth in it (the modern boxes have a lip around them to prevent the bitch from lying on the pups and crushing them). This whelping box was made to match the one used by Jazz, the double, but it was larger in order to give expectant Betsy plenty of room during delivery. Betsy was also given several weeks to familiarize herself with the whelping box prior to her litter's arrival.

The real puppies' birth was filmed on a locked-off camera in private by the production executive Sharon. Animal handler Jill ensured that all the puppies were delivered safely in an environment that preserved their dignity and welfare. Only Jill and Sharon were present, and only Jill was permitted to deliver and touch the puppies. These shots were inserted into the final sequence in the episode with the help of visual effects. The real mum and puppies were never used during the filming with the actors on location. Instead a silicone prosthetic puppy – a 'hero' puppy – was created for Nick to hold and for the dog double to interact with. To complete the litter, prosthetics specialist Pauline made four or five more puppies, moulded from silicone with matching fake hair added with tweezers.

The real puppies' timing was not perfect. On the night they arrived in the world, there was a cast-and-crew screening of episode one of All Creatures, fittingly held at the hotel called Herriots in Skipton; the first opportunity for them to watch the fruits of their labour. The team awaiting the birth couldn't make the screening. In a Portakabin, Jill sat with Sharon, waiting for that moment.

'I'd got the call saying Betsy had gone into her whelping box and wouldn't come out,' says Jill. She knew it would be tonight. 'Sharon and I sat, eating chips and waiting, waiting, while Sharon played the first episode on her laptop. Everyone else was having fun at the screening.' Sharon adds, 'But we got to be at the birth, which is the most privileged and magical experience of all.'

Challenges and headaches, hitches and glitches aside, many of those involved in the series would reflect and say that this was the happiest set they had ever worked on. They didn't grumble and moan, and W. C. Fields' name was never mentioned on-set. Anna Madeley, who plays Mrs Hall, says, 'All of us have very strong instincts on how we want to play our characters, which has been a real joy. It's been a delight all the way through.'

In the early days of the project, the UK was slowly being divided by Brexit. Then, when the series was aired, the nation and the world were united in the struggle against the Covid pandemic. 'These stories bring a sense of hope and connection in a way that's very uplifting,' says Melissa. 'At its heart, it is all about family.'

Family. Two brothers, Siegfried and Tristan, together with James and Mrs Hall, become the most wonderful dysfunctional family, one that we can all relate to. Then there is the family that made *All Creatures Great and Small*. A collection of highly skilled professionals who came together to create the series. Within the family of cast and crew, there are more families including the mother and son animal handlers, Jill and Dean Clark. Mark Atkinson, the horse master, works with his son, Ben. Meanwhile, during lockdown, Alex Harwood, who composed the show's score, enlisted the help of Phoebe Workman, her daughter, who became the music editor for the second series.

Together this one huge 'family' – made up of about seventy people – faces the challenges, Covid included, to make All Creatures for a nation of families to enjoy.

The first episode was aired on 1 September 2020 – five decades since Alf saw the publication of *All Creatures Great and Small*. Overnight millions of us were hooked and 5 million viewers returned every week, for an hour-long escape from the reality of lockdown, to be happily transported to the Dales, to Darrowby and Skeldale House. The public, the critics – they were all wowed. 'The only thing,' says Sam, 'was that somebody wrote to me, saying, "You were all having tea without a tea strainer." I realized – oh yes, in 1937 we were about fifteen years away from having tea bags. I wrote back, "I promise you there will be a tea strainer in series two." For me, that's the big difference between the first and second series. The tea strainer in the second, but not the first.'

The series premiered in America in January 2021. A second series was commissioned although this was to bring its own challenges, as social distancing became a legal requirement and Britain rode the Covid waves. There was regular Covid testing, overseen by the medic on-set, Phil Pease. Cast and crew wore face masks, ate separately and travelled separately. 'There were all sorts of tensions,' says Sam. 'Our extraordinary runner, Gemma – I realized that I had been working with her for thirteen weeks when one lunchtime I saw her eating a sandwich. And I said to her, "Gemma, you know, we've been working together for three months and this is the first time I've seen your mouth."'

On the first day of filming the second series, Sam, once again, took a photograph of the clapperboard, eighteen months after that first picture at the start of filming series one. 'Not only did it have the same director and DP on the board, it was drawn by the same clapper-loader and held by the same camera assistant. And around us were old friends. It's wonderful that so many people wanted to come back.'

The producers could rightly claim a triumph, a phenomenal hit. The series pulled in the highest-ever ratings for an original show on Channel 5. In the run-up to the broadcast, social media had been abuzz, and one of the most frequently asked questions was – will you be using the theme music from the BBC series of the seventies and eighties? The producers had been pondering the question themselves; they could not ignore the subject. Composed by Johnny Pearson, the theme was hummed across Britain and is still well remembered by many.

However, once Alex Harwood had composed a new theme everyone fell in love with it. 'It quickly became the tune that we couldn't stop humming,' says Melissa. 'It felt right for our show.' In a glass-raising gesture to Alex's predecessor, they decided to have Johnny's theme played over the end credits of the first episode of series one. This was a little treat for the audience, once they had been introduced to a new cast. 'It kind of landed everyone back home,' says Melissa.

So how had this new adaptation come about? Who came up with the idea? Who had thought to make a modern-day series of Alf Wight's classic, a series that would have us gathering around the TV, or watching on a laptop in bed? Millions of us have fallen in love with the series, watching with feet up and the cosy-covered teapot within stretching distance. This adaptation, it turns out, was the realization of one man's vision. And strangely, that man's story leads us back to the United States, which was home to St Martin's Press, publishers of *All Creatures Great and Small* …

CHAPTER FOUR

The Idea

'Let me tell you,' says Sir Colin Callender, the CEO of Playground, 'I so love this show and, frankly, I like to produce things that I want to watch. I mean, I don't have any illusions about what the world is really like. We still live in such difficult times and, actually, I wanted to be transported back to Darrowby.'

This is the story of how Colin came up with the idea to transport not only himself back to Herriot's town Darrowby, but also to take millions of viewers on the journey with him. And the roots of the series, *All Creatures Great and Small,* go back to a world and a time far, far away from the Yorkshire Dales of the thirties.

In 2008 Colin was in the States and president of HBO Films, where he had a reputation for producing award-winning, sophisticated television and feature films. The Bush presidency was drawing to an end and, later that same year, Barack Obama would be elected as the forty-fourth President of the United States. 'The country was very divided between the Left and the Right, and it was the beginning of the urban versus the non-urban voting in America, which later would be right at the heart of the Trump election,' says Colin.

With Tom Hanks as executive producer, HBO had acquired the rights to David McCullough's Pulitzer prize-winning book *John Adams*, a biography of the second President of the United States. It was being adapted for a seven-part mini-series, and the story also recounted the relationship between Adams and his political opponent, Thomas Jefferson. 'They had diametrically opposed political views, but in their own way to this day still represent the sort of political divide in America,' says Colin. 'Yet they shared some profoundly common values about America, and about the sort of life that everyday individuals wanted to lead, even if they believed in a different political approach to how to achieve that.

'I remember sitting down at HBO and saying the country is very divided, with the primaries to be held in the spring. If we make *John Adams* and explore this friendship between Adams and Jefferson, which was really where both men shared this great vision of what America could be, we could bring people together through this television event. We could make people who cared deeply about America be together, because they would see in this drama the things that they shared more than the things that divided them.'

His instincts were rewarded. Zoom forward and *John Adams* was a monumental success. The series won thirteen Emmy awards and four Golden Globes – in the history of the Emmys, it holds the record for the most wins for a programme in a single year. 'It won every award under the sun, it got enormous ratings, it got enormous critical acclaim,' says Colin, 'and I genuinely believe that was because, as

I say, it celebrated the things we shared rather than revelling in the things that drove us apart.'

Fast-forward to 2015. Three years earlier Colin had founded Playground, and now his focus was on Britain as well as America. He would see the business flourish to become a leading producer of TV shows, mini-series and films, as well as productions for the theatre. With offices in London and New York, Playground's television productions include adaptations of *Wolf Hall* (starring Mark Rylance, Damian Lewis and Claire Foy), *The Dresser* (with Ian McKellen and Anthony Hopkins), *Howards End* and *Little Women* (like All Creatures, another period classic with a family theme, starring Emily Watson, Michael Gambon and Angela Lansbury).

But in 2015 he took a good look around. 'There was this sense that Britain also was an increasingly divided nation; between rich and poor, urban and rural, old and young, and those who were pro-Europe and against Europe. And certainly in America, it was a similar divide.' (In Britain, the rapidly evolving storm over Brexit – should we or shouldn't we leave the European Union? – would divide the nation almost exactly in two. And in the States Donald Trump would soon become president, carving another gigantic split between the Trump supporters and haters.)

Television drama, too, was following a pattern. 'There was a whole bunch of shows with characters at the centre who were morally ambiguous, and this was particularly true in America with dramas like *House of Cards* and *Breaking Bad*. There was a proliferation of crime dramas with these morally complicated and compromised characters.

I did think that there was a desire among the audience for a return to a kind of gentler time.'

His mind took him back a few years, to that story of America's founding fathers, and he saw parallels between what he felt then and what he was feeling now. 'We needed a drama that audiences could watch together that would embrace and celebrate the values, the views that we shared, rather than the things we didn't. I don't think I would have started the same way with All Creatures had I not had the experience of *John Adams*.'

He did not come up with a shortlist of titles that might meet his criteria for the right kind of drama. Instead he went straight back to *All Creatures Great and Small*. 'It was specifically about remembering the BBC series and remembering how I had watched it with my family – how I felt, and how it made other people feel. There was an opportunity and a gap in the television landscape. I was looking for something that could accomplish it because I felt there was nothing else on television like that at the time. I thought, this will be embraced by an audience; it will fill an audience's unsatisfied appetite at a particular moment in time.'

He reread Herriot's books. 'What I had forgotten was how funny they were and remembering that made me feel, my goodness, we can really entertain while at the same time embrace certain core values that I think everyone shares, whatever their political persuasion.'

Reflecting on the timing of it, he says, 'Then there would be Brexit, and Trump in America and then, of course, Covid. So the accumulative impact of all those things, all that led to a sense that it created an environment so that the audience of the show was even more ready than it was when I first thought about it.'

And the more he considered the idea, the greater his compulsion to act on it. 'I did feel that we could revisit Herriot in a way that would speak to a contemporary audience. One of the early things I thought about – and it's really quite simplistic in some sense – is that the new digital technology would allow us to shoot everything in a way in which the beauty of the landscape and the world of the Yorkshire Dales could be brought to life in a very beautiful way.'

In the early eighties he had worked with a natural history director on *Brendon Chase*, a series of dramas that featured animals at the centre. That decades-old experience inspired him now. 'I thought of being able to shoot the beauty of the Dales with the animals, and dramatize this in a way that captured the humour and developed the female characters more fully and gave all the characters some real emotional and psychological depth. I thought we could bring this back to life and entertain in the way that the BBC series had done, but for a contemporary audience.'

One morning in the autumn of 2015, Colin left his office in Drury Lane in Covent Garden and made his way to the office of David Higham Associates, the literary agents in Noel Street, a mile away and bang in the heart of Soho, London, a quick stroll from the hustle and bustle of Oxford Circus. With its busy streets, tourists and shoppers, and the cacophony of car horns and sirens, this part of England is starkly different to the Dales, about 225 miles to the north.

There the towns and villages are separated by a seemingly endless, rolling tapestry of fields, limestone cliffs and grassy hills, laced with gently flowing streams and lined with ancient drystone walls and, all around, grazing sheep, horses and cattle. Jim and Rosie – son and daughter of Alf Wight – had travelled from Thirsk to the capital to have a meeting with Colin. They were joined by Jim's daughter, Zoë. They sat with Colin and listened as he told them of his dream, to adapt their father's books for a new television series.

In Jim's biography of his father, *The Real James Herriot*, he describes how Joan Wight had encouraged her husband to send his manuscript to David Higham. A few weeks later Jim saw his father at the breakfast table, 'his hands trembling as he fingered a letter that had just arrived. The letter was from David Bolt, a director at the company, saying that he liked the book "enormously" and considered it would have every chance of reaching publication.'

'Jim, I can't believe it … my book might be published,' Alf said. 'After all these years! I just can't believe it.'

Jim must have remembered that as he and his sister sat there now, listening to Colin's hopes of adapting the book into a series that would be screened on prime-time British television and in the States. Colin remembers: 'What sticks in my mind is that they were immensely protective of their father's legacy and their father's work.' He was not deterred, however. 'I wanted to reassure Jim and Rosie that I was eager to embrace the core values and the tone, and essence of the work.' He adds, 'I think they thought that they were scaring me. Actually, I was very pleased about that because I thought they could help us. They would be able to help us to get it right.'

Colin, Jim, Rosie and Zoë had first met to discuss the project on 16 September 2015. It was forty-four years – less a day – since Alf Wight signed that contract with his US publishers, sealing the deal for the publication of *All Creatures Great and Small*. And, though they could never have imagined it, precisely four years to the day after that meeting, the cast and crew would assemble in the Dales, filming two memorable scenes: James, naked and swimming beneath a waterfall in the icy lake (he's surprised when Helen appears); and James driving through the Dales, wind in his hair, Siegfried beside him in the passenger seat.

Coincidentally, episode one of the second series was aired on Channel 5 on 16 September 2021, five years to the day after Colin's first meeting with Alf's children. A special date indeed in the Callender calendar.

Truth be known, Colin was not entirely confident that All Creatures would be a hit. 'In fairness, it wasn't a definite that we could bring the show back and would be able to emerge from the long shadow of the BBC series,' he says. 'That wasn't a sure thing. But I think we have, and it's clearly found a whole new audience and I do think in a rather sort of magical way it is speaking to a particular moment in time.'

He had seen the rushes every day during filming. 'I have to say that the real triumph is the work that Brian did. His casting of the series was utterly impeccable. You know one of the challenges was Siegfried. Robert Hardy's performance is somewhat iconic and that was a tough act to follow. I think my early experience of watching the rushes was seeing Nicholas and Callum and all these actors inhabit these roles and, within seconds of seeing them do so in the rushes, believing that they *were* the characters. I think that was my overwhelming response initially.

'As much as I loved the BBC series, I didn't think that ours was somehow a pale imitation. It wasn't trying to replicate that series … And again this is a tribute to Brian Percival because the actors almost sort of instantaneously inhabited the roles and made them their own, and I just got caught up in it. And for me, watching it as a producer, that is when I saw that and felt that about all of the actors. That is when I thought, OK, we have a real chance of this show working.

'I felt that the actors were bringing something fresh and new to the screen around the characters that sat side by side, honourably, next to the BBC casting, but it *was* different. And I loved the idea of discovering new actors with Rachel and Nicholas and Callum, on the one hand, and then bringing on board actors with great resumés like Sam and Anna, and then, of course, some of the cameo roles with Dame Diana Rigg and Nigel Havers …'

He was at home in New York when he watched the first episode. How did it feel to see the dream become reality? 'It was a joy and far exceeded my expectations.' But he wasn't that sure the show would bring families together in front of the telly, as it had done with the BBC series. 'Today there isn't just one TV set in the house. There are either multiple television sets or people are watching on their computer, laptop or phone. So it wasn't clear to me that the ambition of getting families together was necessarily realizable.'

The Covid lockdown did bring families back together, though. 'There were so many emails or texts or social media posts with grandparents saying, "I watched this with my grandkids," or people saying "I watched this with my parents." So clearly that did happen, but I think that was partly to do with the nature of the show and partly to do with lockdown. But I do think the central notion, that at its best television drama can bring people together, was at the heart of this. The reason the books are so beloved is that they appeal across the board.'

Melissa Gallant recalls watching the first episode as 'quite a moment for all of us'. 'A first cut of a new series doesn't always work well,' she says. 'Plenty of great shows begin with a cut that doesn't quite work yet. But it wasn't the case with All Creatures. Even though we'd seen every minute of the rushes, watching it as an episode for the first time is entirely different. Richard and I watched the episode in Skipton with Brian and the editor, John Wilson.' They were up in the production office, what was once a headmaster's office when the building had been a school. The room was now the temporary edit suite.

They watched mesmerised, chuckling, smiling, invested in a story that they knew so well. 'The quality of the cinematography and performances, the look and feel of the show was stunning. Brian and the team had created a world that we didn't want to leave and delivered a show of the highest quality. Nicholas was faultless, Yorkshire looked breathtaking, and the show evoked a range of emotions that we all wanted to feel again and again.'

They thought that they might just have something special on their hands, 'but you always have to be careful not to fall too much in love with your own show because you're making it for the audience'. Colin had watched the link and the minute it ended, he phoned Melissa from New York. 'I happened to be in the supermarket when I took his call. He had the same reaction as we had. He loved it.'

Melissa invited Jim, his wife Gill and Rosie to Skipton to show them episode one on a big screen. 'It was like being on your driving test, when you're constantly wondering what they are thinking. There were chuckles in the right places, Jim seemed happy with the veterinary procedures, and they enjoyed recognizing the locations.' They watched all the other episodes via a link on their home computer. Jim's verdict on the adaptation of his father's books: 'Brilliantly done, brilliantly produced, brilliantly acted. It captures the Dales, the era, the spirit … And Sam looks like Donald.'

CHAPTER FIVE

Faith

When Ben Vanstone was a boy, eight or nine years old, he'd sit enchanted as he and his family watched the BBC's adaptation of *All Creatures Great and Small*. 'I suppose it was one of the first shows that I remember as a kid,' he recalls. 'Watching with my family on a Sunday night.'

Today he is the lead writer on the series, and a writer with a regimented routine. 'I tend to keep the same schedules. I get up at the same time every day and I cycle to my office. I go to the gym and then I'll start work, and then I'll stop and have lunch at exactly the same time every day.' Often he writes listening to classical music: 'I try to play music to match the moods of what I'm writing. Nothing with lyrics – lyrics completely throw me.'

Much of his thinking – the contemplation – is done before his fingers touch the keyboard. 'So when I write I'm just in the moment, seeing the characters in my head, writing down what's happening rather than trying to think who says what next. I'm noting down a finished product that I'm watching, and not trying to think what's happening.'

Often, and especially when he is on a tight deadline, he dreams about the scenes. 'I'll wake up and have them in my head. I jot them down very, very quickly.' Splitting the time between writing and thinking, writing is 30 per cent of the job, while actually thinking about what to write accounts for 70 per cent. 'I try to allow myself to step away from the keyboard so I'm not always being confronted by a blank page. If it's not flowing it is always better to take a step back and think about why, rather than trying to force yourself to write.'

Ben recalls that initial meeting with Colin at which they discussed the new adaptation of All Creatures. 'I was really curious what we would be trying to do, and I was a little bit trepidatious because this was such a big show. I wasn't quite sure what we'd do with it, and how it would be different and earn its own place. It was explained that they wanted to do their own new version, a more cinematic one.'

He had read the Herriot books before the meeting and was excited to see 'a whole layer' that seemed unexplored. 'I got very excited about that. I thought there is light and dark, and a deeper emotional level that I felt wasn't quite fully explored in that previous series. I wanted to find a way to access those stories and that part of Herriot's world, and bring it to the screen. For instance, there was more reflectiveness of James and his outlook on the world. And in the books there are hints that there is more to Siegfried's madness than simply him coming into the room and shooting his mouth off at James and Tristan. The books raised so many questions about this character.'

This was not the first time that he had adapted one of the classics. He wrote the screenplays for *The Borrowers* (starring Christopher Eccleston and Sharon Horgan) and *Cider with Rosie* (with Samantha Morton). However, he had never written so many animal stories and was yet to discover the complications of filming with them. The things you don't think about. 'You write there's going to be a pig charging at Tristan and then everyone gets the script and panics – "Pigs can be heavy and we don't want to injure Callum." Then you discover the government stipulations regarding the movement of animals and everyone curses you for it!'

He set to work on the story of a young man who follows his passion. After graduating from veterinary college, James leaves home in Glasgow. This is 1937, and there is little work available. We sense the desperation of James as he arrives for his interview in the Dales. Siegfried Farnon (based on Donald Sinclair) is unpredictable, forthright and obstinate but hilarious. This eccentric will become James's boss. And James has plenty on his plate. There is the job itself, and the difficulties of dealing with straight-talking farmers. And even some romance. James falls in love with Helen, the character who so intrigued Alf's publishers in New York.

With Colin and Katie Draper, Playground's then co-creative director, Ben developed the script for a pilot show. It was sent to Ben Frow, director of programmes at Channel 5. He loved it at first sight. After discussions, in January 2020 All Creatures was 'greenlit', industry speak for getting the thumbs-up and some cash to go ahead.

Later, when Ben Frow watched an early cut of episode one, he wanted instantly to greenlight a second series. This is unheard of. Second series are usually greenlit when enough of a show has transmitted for them to be confident about the audience figures. Commissioners have been known to give a greenlight when a whole series has been made, with perhaps just one episode aired. All Creatures hadn't even finished filming at this point, and the episode hadn't been properly graded or sound-mixed. But Ben Frow loved what he saw and he wanted that next series.

Ben Vanstone and the team had also written a 'series bible' for All Creatures. This evolved as the project progressed. Their bible got bigger. It is a document which maps out the characters and the development of the series, episode by episode. It is an invaluable reference point for the writers, directors, producers, cast and crew. Every aspect of the show is covered. There are sections devoted to the look and the feel of the show, analysis and exploration of the characters and their stories, and there is a section entitled The Herriot Estate. It reads:

> They have a wealth of memories, insights and knowledge about their father and his books, and of course invaluable historical veterinary expertise. Jim Wight is the author of *The Real James Herriot: The Authorised Biography*. Jim Wight and Rosie Page are key to the process through both scripts and production and are a vital resource for us to draw upon. Their input will enhance all aspects of the series and help to ensure that our adaptation captures the authenticity, tone and accuracy of James Herriot's world and stories.
>
> One of Herriot's grandchildren, Zoë Wight, works in the filming industry as an art director. She has also been incredibly

helpful to the development process and her industry expertise means that she is well placed to advise the family on the process of modern TV drama production.

The bible outlines the arcs of the story, development of the characters in terms of their emotional journeys and their character stories. Ben also had broad ideas of what animal stories they might use to access those character journeys. And just like the Bible, this bible instils the reader with faith; faith in the project because it shows that Ben and the team have a clear idea of the way ahead.

Melissa, who joined the team as executive producer, had a background as a script editor, and she worked closely with Ben. They didn't have long to find the right format for the series and develop the rest of the scripts. 'Ben had completely immersed himself in Herriot's stories,' she recalls. They set the show within the three worlds that feature in the books: the farms that nestle in the Dales' landscape, the fictional town of Darrowby, and the professional and domestic world of Skeldale House.

'Ben told stories that weaved in and out of those three worlds that explored the reality of being at the behest of nature and stories about an unconventional family that were very relatable. Ben is brilliant at capturing the tone and spirit of Herriot's books … Herriot will tell a story that's rather poignant and sad, and have you laughing the next minute. Part of the key to capturing Herriot's world is to find those shifts within a story, a scene or even a line between something that's truthful and painful, and something that's funny. It plays with such ease on screen, but it's incredibly difficult to write, direct and perform – it can't ever be too earnest nor too silly. The tone of the show is incredibly specific and if it's just a few degrees out it doesn't work. And it all comes from character.'

Four more writers came on board: Lisa Holdsworth, Freddy Syborn, Debbie O'Malley and Julian Jones. The bible was sent to them before they gathered. 'Everyone looked at that document and bashed out the odd email and read the books,' says Ben, 'and then we all came together for a week.'

The writers came together at Playground's London offices in March 2020, six months before filming began, along with Melissa and Ciara McIlvenny, the script executive. This was their writers' room. Ben describes this creative bubble: 'There is no head of the table, as it were. People can sit where they like, eating nuts, sweets and biscuits, and whatever else to stop them falling asleep. I tend to wander around the room because I'm not very good at thinking if I'm sat still. I'm a bit of a pacer and a wanderer, and we'd start by talking about the characters.

'Our first port of call was to discuss James first and foremost, and then Siegfried and Tristan, and Helen and Mrs Hall. We worked our way through those characters, making notes on a big board on a wall, and starting to think about how those stories would be plotted across each episode. For instance, with episode seven, we always knew that we were ending with the wedding and James is going to be with Helen up in the high Dales. Does he tell Helen how he feels? Or is he actually too much of a gentleman for that? All of this was worked out and worked through before we started writing everything down.'

Then they focused on achieving their aims in the individual episodes. After a week in the writers' room, they went away and wrote bullet-point suggestions for each episode, which they emailed to each other. 'Then we all looked at the points – "That's not quite right," "That doesn't quite work," or "That's great!"'

And so they began …

One day Sharon Moran was strolling along Shaftesbury Avenue in London's West End when she bumped into Richard Burrell. They hadn't seen each other for a couple of decades and, after a cheerful chat, they went their separate ways. Several months passed when …

Noëlette Buckley, Playground's head of production, Melissa and Sharon started to think about producers and directors for the series. 'I mentioned Richard's name,' says Sharon, 'and Melissa said, "I know Richard – I've worked with him."' In 2006 Richard and Melissa worked together on *Robin Hood*, the BBC series. Would Richard be interested in teaming up again, on *All Creatures Great and Small*? 'It was a no brainer,' he says. 'I put my hands up and said I'd never read the books. But they were still a part of my childhood in that my parents read them, and my sisters, and then from 1978 the show was on.' In fact, one of his sisters had a crush on Peter Davison, who played Tristan in the BBC series. 'She got to meet him at a fete and shook his hand,' he adds. 'Don't think she washed that hand for a week.'

While the team developed, so too did that series bible and the stories which, one day, would make it to our screens. Through Ben's scripts the presence of Mrs Hall was elevated, from being in the background to firmly foreground. Colin Callender says, 'Mrs Hall would be a crucial character in this adaptation. She becomes an emotional pivot. These three guys who are flailing and running in opposite directions – she keeps them all grounded and honest. And it's a wonderful role played by Anna Madeley.' The series bible set out the character of Mrs

Hall as 'loyal … a pillar of the community and a good Christian, and the moral centre of our show'.

These days the main cast members contribute to the process of character development. 'There was a conversation I had with Anna and Rachel. They were trying to work out how well their characters knew each other – whether Helen would call her Mrs Hall, or would she use her first name, Audrey. It seems like a small thing but it's quite important because it tells us everything about how well they know each other. That made me think about it and we discussed it, talking through the back stories of those characters. We landed on her being called Mrs Hall by Helen and then that evolves in series two, with Mrs Hall becoming Audrey to Helen. It's only happened through that sort of collaboration with the actors.'

There was, they all agreed, one big star in this series and it had neither an agent nor an animal handler and required no vet. It was Yorkshire. Part of the joy of the series would be, quite simply, the beauty of Yorkshire and the Dales – 'The whole world,' says Colin, 'in which these stories are set.' The series would present an authentic, realistic picture of rural England in 1937.

Richard adds, 'This is a time when many of the farms out in the Dales would not have had electricity. They'd be smallholdings, perhaps with a small flock of sheep or a small herd of cattle, and for them the role of a vet was key if their cow was to make it through the night, or a calf be born successfully. We wanted to show that world.'

In February 2020, a month after the greenlight from Channel 5, Melissa and Katie went to Thirsk to meet Jim, Rosie and Zoë. 'We talked about how they would be the gatekeepers to their father's books, legacy and memory,' says Melissa, 'and I would be the gatekeeper to the show. And we would work together to get it right for the audience, and for them.' On her next visit, Melissa took Ciara to meet them. 'Next we went through episode one, page by page so that we could hear every thought and comment that they had. And Zoë joined us with her eight-week old baby, Alf Wight's great-grandson.'

Melissa kept them up to speed with all developments, such as the main creative appointments, casting and scripts. Jim became a consultant, offering advice and wisdom on the veterinary procedures in the scripts. He and Rosie continue to see the scripts in advance, making suggestions and spotting potential veterinary howlers. Says Jim, 'We always find something wrong, but very often it's nothing earth-shattering and as time has gone on there's been less and less.' Zoë, who is a television art director, helped to answer any queries that her father and Rosie had about the programme-making.

Melissa came to regard their show almost as a palimpsest: a scroll or manuscript which has been scraped and cleaned so that, once more, it can be written on or inscribed, bearing visible traces of what has gone before. 'With All Creatures there are several layers; the real Alf Wight; the fictionalization of his life in the Herriot books; then Jim and Rosie's recollections of their father; and then the BBC series. Our adaptation is a sum of all those layers, plus a new contemporary exploration of those characters and their world.'

All Creatures was fast on its way to what Melissa describes as 'a passion project'. 'Herriot's stories are joyful. They make you feel better,' she says. 'And everyone who came on board felt that way about the material. They loved Herriot's world, Ben's script, the Yorkshire Dales. We kept having people join the team who somehow weren't supposed to be available. Quality attracts quality.'

Everyone was inspired by the quality of Ben's scripts, and Colin says, 'Ben made the point early on that there are no villains in *All Creatures Great and Small*, which I thought was such an astute observation. There are no murders, there are no carjackings, there are no big action moments.'

Colin mentions one of his favourite scenes in series one, episode seven, the Christmas special. It's Christmas Eve and Mrs Hall has gone to midnight Mass but her son hasn't turned up as she hoped. Siegfried arrives at the church, slightly late, and stands by her. Colin says, 'He notices she's crying and he puts his hand on hers. That is what is at the heart of All Creatures. It is just little moments like that.

'I have never met a murderer so I don't know how I would deal with it. But I have certainly had a friend stand next to me crying, and I've put my arm around them. Those are the moments we know, can understand and relate to. Those opportune moments have helped to make the show a success.'

PART TWO
Setting the Scene

'I love it here; the work, the place, the people.'

– JAMES

CHAPTER SIX

Picture Perfect

A n old but well-kept railway station in the Yorkshire countryside. It is night-time and raining. In the dim light we see a man – trilby, overcoat – waiting on the platform. Next, a steam train pulls into the station, braking wheels grating on the track. A woman steps from the train, a cloud of steam trailing in the background. Close your eyes and you can almost smell the smoking coal of the engine. The man, the woman, they are in love. And as they meet a tear runs down the man's cheek. She takes a tissue and gently wipes it away. Just one of those small moments we can all relate to …

'Cut!'

A young Brian Percival, a student of film at Manchester Polytechnic, was the director. This was a commercial for Scotties tissues. Passionate and driven, Brian was making it off his own bat. The lovers, a steam train, all set to the music of Rachmaninov. This short piece of film evoked the spirit of that beautiful black-and-white classic, *Brief*

Encounter. Who knew that many years later Brian would be filming on the same railway line for *All Creatures Great and Small*?

Looking back, Brian says, 'I'd gone to – of all places – the North Yorkshire moors and said I'd love to make this ad. Showed them all the storyboards and stuff like that. And they went, "Well, yeah, OK." But the trouble was, it was being shot at night. So I had to somehow source lamps and generators and a crew for the lighting. Then I went to a lighting company in Manchester and said, "If I get the train, will you light it?" And they said, "Yeah."' The commercial won him a Kodak Award.

In the late spring of 2019, Brian was again in the Yorkshire Dales. A couple of decades had passed since he had last been here. He and this northern region of England may have been separated far longer had it not been for a recent call from his agent. 'I don't know if this is for you, Brian, but I just thought you might like to have a look at it…' Brian did indeed want to have a look, and soon afterwards he received a draft of Ben's script for *All Creatures Great and Small*.

He was captivated. The story itself, the way it was told, were different to the BBC series that he had watched as a teenager in the late seventies. 'Ben's scripts were lovely and made me realize that the thing that was missing from the previous series and the Herriot books – there weren't any strong female characters. Ben had addressed that issue. Helen had come more to the fore, and Mrs Hall was a main character.' The sheer escapism of All Creatures also appealed to the director. 'I've always enjoyed making films and programmes that other people can enjoy,' he says. 'I've never seen anything wrong in doing something that is popular, and that people can enjoy – escape their lives for an hour. I think people need escapism.

'Over the years I've done things that very few people might see but it might get some critical acclaim and be quite powerful. Equally, I like the idea of people tuning in on a Sunday night and seeing something that gives them a chance to escape.' Brian wanted his own form of escapism: he was looking for excitement, a change to his life. 'About six months earlier I'd had heart surgery and had taken a few months off. I was fine and ready to go again, and looking around for something to do. Something new.'

An interview was arranged, at which Brian would discuss All Creatures with Melissa Gallant and Richard Burrell. He was one of a number of directors on the Gallant-Burrell shortlist, but the quality of his work and his track record put him very high up. However, before the interview he happened to have a commitment in Yorkshire, and decided to drive up through the Dales. Avoiding the main roads meant a longer journey, but he was keen to reacquaint himself with the national park. This drive was not just curiosity, it was a 'recce' – a reconnaissance mission to check what he might be up against should he become the director. 'I didn't think about it at the time, but I drove down the road that completely by chance we've now driven down about a thousand times. It's right in the middle of where we film everything.'

Surrounded by the breathtaking view – majestic, imposing, arresting – he felt an escapism which he strongly believed audiences wanted, required and would appreciate. On the car radio, news bulletins covered the latest developments in the lengthy negotiations between the British government and the European Union. The nation was exhausted by the relentless political drama. 'It was Brexit, Brexit, Brexit, left, right and centre. Just like many others, I was so sick of it. Wouldn't it be lovely to have this escapism,' I thought. He'd turned off the news and listened to the audiobook of All Creatures.

Then he pulled up, got out and took a long look onto the valley before him. He was not auditioning an actor but he may well have been at that very moment because, as he absorbed the view, he saw the Dales as a character in the series. Just as any character develops within a story, so too can the landscape, and Brian is a keen landscape photographer. 'You have to accept that beauty comes in different forms,' he says. 'It can't always be with sunshine. Sometimes you look at the moors covered with mist and it's pouring with rain and it's stunning. It's a different sort of beauty. That's the whole thing about landscapes – part of it stays the same, but much of it changes. It's all about embracing that.' He was thinking about Yorkshire in the same way that Colin Callender had done, but they had yet to discuss this project.

The ever-changing landscape of the Dales had him hooked. 'I thought to myself, "I can do something with this."' He continued his drive and, a week or so later, met Melissa and Richard. Brian had now signed up as director, who had not only read and loved the script, but was passionate and excited. He had a vision. He is a master at building 'worlds' on-screen.

Melissa and Richard sat, almost mesmerized, as Brian recounted how (as director) he had envisaged the world of Downton, from the set to the sound design, both above and below stairs. He would bring that same creativity and artistry to All Creatures, in collaboration with Jackie Smith, the show's production designer.

There were other fundamental aspects of the story that presented parallels of Brian's life. For instance, his own early years were echoed in Ben's script: the opening scenes show James Herriot in Glasgow, the son of a docker, and establishing a way to leave the city for a better life. Brian, too, has working-class roots, and his father worked on the docks in Liverpool. 'A lot of coal that was exported made its way from the collieries in Yorkshire and the north of England, and would go out through Liverpool. The coal wagons would come into Liverpool docks, and they rolled up the piers and tipped the load into the ships. My dad was in charge of checking what went where, what been tipped or what had gone missing.

'At the end of our little terraced street, there was a wall, then a railway track and then the docks.' He pauses and then adds, 'There you go, railways and steam trains – they were there in my childhood and they've featured in much of my work.

'In his spare time my dad did a bit of painting and decorating. He was a very proud man and, typically of the time, we never knew how little he was earning until he died and we found a wage packet. That was just the way things were then, in the sixties and seventies.'

Among his father's few cherished possessions there was a camera. As a lad, Brian and his family used to have seaside holidays in Meols, a village on the Wirral Peninsula, not far from their home in Liverpool. 'My dad used to take holiday snaps and then he'd show us the pictures on slides. So I was always interested in photography.' And is that what prompted his career? 'You know, it could well have been. That's maybe when I was first aware of someone taking photographs, and it is a very strong childhood memory. And he left his camera behind. Not that I ever took any photographs with it.'

Richard Burrell reflects on that meeting with Brian and says, 'We'd met many talented directors, but once we sat down with Brian it was almost like what we wanted for the show was coming home. It was not an interview. It was three people – Brian, Melissa and myself – chatting in a room, and suddenly it all seemed so easy and everyone was on the same page and in accord. This was the director who could help us create that world.'

Melissa recalls that initially they didn't think Brian would be available. 'We were on tenterhooks waiting for an answer but unbeknown to us, Brian had been driving around the Dales listening to the Herriot audiobooks, falling in love with the landscape. And he'd been very taken with Ben's script. Brian said after the meeting, "I could have talked to you two for hours." We pinched ourselves when we heard he wanted to do it.'

As for Brian: 'The best thing I ever did was to say yes.'

One morning, Richard Burrell set off for Yorkshire. The drive gave the producer a few hours to reflect on the project. He'd loved reading Ben's script. The Yorkshire countryside itself was to be one of the characters and not merely a backdrop, and that was what brought him here on this particular day. Actor or landscape, when either is great it's compelling to watch. Today Yorkshire was auditioning for Richard. He'd be scouting for potential locations that could be used for scenes in the production of *All Creatures Great and Small*.

His plan? 'You have a map and you think, "Where are we going to be?" We decided on the market town of Skipton. That was our base from where we ventured out into the Dales to find the locations.' We needed to think, too, of the practicalities of the locations. 'What we do is show business. The show side is the exciting side and the creative side. The business side means you have to make the show in a practical sense and for a certain amount of money.'

At the back of his mind, Richard knew that he and Melissa would need to address the process of filling the key behind-the-scenes roles for the series; apart from the animal experts, they'd have to appoint, for instance, a production designer, costume designer, hair and make-up artists.

For now, however, the mission was location, location, location. As he drove through the Dales, he was so taken by the scenery that he stopped for a moment, took his phone from his pocket, held it up to the vast expanse before him and – snap! – photographed the stunning landscape. He posted the shot on Facebook. Then off to see more locations.

Posting that picture took just a moment, but it was a shot that would bring an unexpected heap of good fortune, for him, for the production of *All Creatures Great and Small* and, ultimately, for you, me and the millions of other viewers at home.

CHAPTER SEVEN

The House that Jackie Built

Jackie Smith was teaching an A-level art class at Skipton Girls' High School. It had been eight years since she decided to take a break from the film industry, and its demands that meant often working away from home. 'I had wanted to focus on my family and bringing up my kids,' she says.

And on this particular day, during a break between classes, she glanced at her Facebook feed. 'I saw some of Richard's beautiful photos of the Dales.' I thought, "Oh, he must be on holiday up here … I'll get in touch and see if he wants to meet up."' They had first met back in 2008. She was the production designer on *Filth: The Mary Whitehouse Story*, the film produced by Richard Burrell. However, they hadn't seen or spoken to each other for years.

So Jackie sent Richard a message: 'What are you up to?' He responded by saying that he was looking for locations. 'I might be able to help,' texted Jackie. 'Do you need a school to shoot in?' Richard: 'No, I don't need a school. But I do need a production designer. Interested?'

She was – she is – a huge fan of Yorkshire's best-known vet. As a child growing up in Yorkshire, Jackie had read every single one of the Herriot books. From its initial broadcast in the late seventies, the BBC series was also essential family viewing in her home. Jackie says, 'Needless to say, I bit his hand off.'

'Richard knew I was a Yorkshire lass,' says Jackie, 'so it seemed I may be a good choice …' She also knew Brian, having worked as a production designer on his 2003 film, *Pleasureland.* 'There were a few meetings to go through and I created a sort of look-book for the show. I met Brian in a Starbucks halfway between Liverpool and Skipton to make sure he was happy with me joining the project. From then on I didn't look back.'

She found it sad to say goodbye to her students, 'but the silver lining is that I have had the pleasure of offering some of my talented alumni jobs in the art department with us. I grabbed Annabel Bower after she graduated in architecture and Sarah Akbar after she graduated in special effects, both super-talented girls who I had taught since they were fifteen.' That felt good she says, 'and I continue to try to bring in students whom I've taught and give young people a chance to work in this exciting industry'.

The Old Mill Becomes Our Studio

There was an important decision to make: should they build a studio set or should they rely instead on locations? Jackie had lots of discussions among the team. Noëlette Buckley also played a crucial role. Behind every successful television production is a highly skilled head of production. Playground's Noëlette works tirelessly to drive the entire production from initial budgeting and scheduling, through every twist and turn to final delivery of the programme to the broadcaster. Noëlette was fundamental to the entire planning and setting up of the show and ensured its safe passage and landing. She also happens to be an animal lover who owns a dog, a cat, ponies, chickens and several cows named after Hollywood film stars, such as Rita Hayworth.

'There were many, many conversations,' recalls Jackie. 'With it being a first series, we weren't sure whether it was the right way to go, but there are many benefits of having a set rather than a location...

'One benefit is that you can "float" walls – you can exaggerate the sizes of spaces in order to make them easier to work in. Another is that camera traps can be created, giving complete artistic control over what's on camera.' Locations can also present challenges. 'The owner might have certain stipulations as to what you can and can't do. In the studio we have carte blanche. As long as the decisions I make are in keeping with the creative vision for the show, then it's all good.'

And so a large, old mill building was acquired, and in it the studio set – Skeldale House – was created. This was a real challenge because it was spring 2019 and Brexit was looming. Warehouses were being snapped up for storage and stockpiling, and were not in ready supply. 'Finding somewhere available, big enough, that it provided what we needed was a tall order,' says Jackie.

Eventually they acquired a warehouse near Harrogate (a thirty-minute drive from Grassington, the village that was to be their Darrowby). This was excellent news for the camera and lighting departments. Everything in the studio set was considered with the film-making in mind, from where to put cables using mouseholes, and equipment, making corridors and doorways wide enough to accommodate a camera track, and ensuring sight lines were lined up to enable long, layered vistas through the set.

From the moment she was taken on, Jackie threw herself into researching Herriot and the Yorkshire of the thirties. She visited Nidderdale Museum in Pateley Bridge and went to a small museum in Grassington, discovering little-seen photographic archives. She trawled the internet. Talked to the locals. A trip to The World of James Herriot, in Thirsk, was highly rewarding. The museum occupies 23 Kirkgate – once the Sinclair-Wight Veterinary Practice, where Alf Wight had worked and, as a young vet, lived.

It was Alf's inspiration for Skeldale House and now, decades later, Jackie considered it crucial that the studio set should pay homage to this same Georgian townhouse. 'The house in Thirsk has a long, skinny, labyrinthine layout,' she says. 'From the front door the hallway snakes through the house, leading you eventually to the scullery at the back. The corridor forms a spine for the whole house, and you feel that you are stepping into the everyday lives of James, Siegfried, Tristan and Mrs Hall.'

She also used some of the original architectural features of 23 Kirkgate. 'For example, the geometric-tiled hallway makes a nod to the original house in Thirsk, and it leads the eye through the space. It also brings the paintings of the Dutch Masters to mind, those celebrations of domestic life with the chequerboard floors, which lead the eye into the space.

'We used salvaged fire surrounds and doors, which had already had a life, rather than making everything from scratch. It wasn't about copying the original house; more about finding pieces that had a similar flavour so that the tone was homely and grounded in truth. All those people who have visited that museum over the years, and the Wight family themselves, felt that our studio set had an honesty and truth to it. So I feel we achieved the right balance.' Indeed, Jim Wight says, 'When I visited the set I felt as if I was stepping back in time. It's brilliant.' In the series the phone number is 'Darrowby 2297'. This is on the phone in the hallway of the studio set, and was the telephone number of the veterinary practice in Kirkgate. It's also the number on the phone in the hallways of The World of James Herriot. The producers used it (with Jim and his sister Rosie's blessing) as another nod to authenticity.

Jackie collaborated closely with the lead director, Brian. 'We talked about the shots he wanted to achieve, about having loads of depth in the studio, so that when you're standing at the front of the house in the hallway you can actually see all the way to the back through a series of doorways and windows. Brian is an artist himself so, from the production designer's point of view, he's a wonderful director to collaborate with. He absolutely gets the whole creative side of things. He wants to talk about colour and textures, and about use of space. And you can't ask for more than that.'

Jackie was inspired by the paintings of seventeenth-century artist Pieter de Hooch, which show domestic interiors, with one room leading into another room, and hints of domestic life going on in the background. Jackie set out to emulate this in the set, 'so that when we are in the dining room we can also see behind into Siegfried's office, and into the sitting room and the hallway. There's always a view through from anywhere. This gives character and depth to the spaces.'

As with everyone who had read Ben's scripts, Jackie was hooked by the writing. 'All the good parts of the books were in those scripts, but there was also a depth to the characters which enhanced them, and made you feel like you were part of this family.' The production design, she says, always starts from the script. 'Along with the collaboration with the director and producer, it is the script that informs all the design decisions throughout the show.'

Brian talked of creating a 'playground' for the cast and directors. This would provide numerous possibilities and lots of camera angles, with interlinking spaces that could be closed off or walked through. He wanted to ensure that they had not used up all the possible camera angles by episode three, so that other directors could find other ways of using the spaces and set up different shots. And that's exactly what happened.

Detail, Detail, Detail

In order to bring the scripts to life – quite quickly – Jackie had to almost immerse herself in a certain time, a part of history, a place. 'You become a mini-expert in a period,' she says of her job. 'Here we are in 1937, and so every prop in this shoot has been vetted to make sure that it fits the period.'

She found some original wallpaper from the thirties, perfect for the walls of the scullery. 'The pattern brought to mind drystone walls and has a floral motif. To me, it was a metaphor for what we were trying to achieve. An underlying structure of strength and grit, but with a kind of layer of chaos over it. This could be seen metaphorically in terms of the grit shown by the characters in the story and the chaos coming from the everyday dramas they encounter, and also literally as it references the landscape.' It wasn't simply old wallpaper, but symbolic. 'It summed up what I was trying to achieve with the set and really the whole show. It portrays how nature disregards our attempts to try to organize it.'

She was influenced, too, by the landscape paintings of a local artist, Simon Palmer. 'They're in the ilk of Stanley Spencer, and Simon often uses little pops of red in a kind of green and brown vista and that draws your eye.' She looked at the works of photographer William Eggleston, 'who also does a similar thing with the red, and we tried to do that with the landscape using, say, a letter box or a phone box, and with the costumes'.

These techniques are apparent, of course, but subtle. In the first episode of series one, for instance, Helen is seen in the vast green landscape wearing a red coat. 'That red coat immediately draws your eye,' says Jackie. 'And Heston Grange, where Helen Alderson lives, is beautiful, with that lovely pop of red from a little letter box before a bridge, and then this glorious Georgian farmhouse set in the Dales.'

On her daily drives from home to work and back again, whizzing through the countryside, she listened to audiobooks of Herriot, being entertained by the vet's adventures as she took in the same magnificent views that had captivated Alf Wight. 'Hearing the words of James Herriot in my ear obviously had an influence on the work. But what a marvellous commute!'

However, filming in a rural location rather than a city, or very close to one, can have its drawbacks. 'It's more difficult to get props, for instance. Everything has to be brought in. But that's why we have tried really hard to source all the props locally from auctions and junk shops, so it doesn't just all come from prop houses in London. There's an eclectic feel to the design.'

Each and every little detail was carefully considered, and Jackie was meticulous as, bit by bit, she brought the set to life. In Siegfried's examination room, his desk is chaotic, while his operating table is spotless – he is disorganized with admin papers, but laser sharp when it comes to the treatment of animals. Texture is added to the overall effect by the hallway's soft-coloured wallpaper and lincrusta, a traditional embossed wallpaper that was typical of late Victorian and Edwardian homes. 'It's about pattern layered across pattern, but never jarring, so that behind the actors' heads the background is soft and not harsh. That way we are focusing on the actors.'

Locations

Jackie, Brian and the location manager, Gary Barnes, drove hundreds of miles around the Dales, looking at farms as they tried to link a location to a reference in the script. 'We have to take into consideration the way that the actors need to move in the spaces, and of course the location needs to work for the story. Equally, we're trying to showcase the Yorkshire landscape. We're trying to show the remoteness of these farms and how difficult it was in those days for vets to get to these places. Some of them are quite inaccessible and we wanted to show the reality, the grit, of a certain era.'

'I Feel Like I Live Here'

From an early stage, Jackie also worked closely with costume designer Ros Little. One of their initial conversations focused on the colour palette that would be used for the studio set design and the costumes.

'Skeldale House,' Jackie says, 'has a kind of buttery warmth to it, almost inspired by home baking. I used an analogy of a fat rascal, which is a large Yorkshire scone made with currants and cherries, almonds on top. It's utterly delicious and served with butter and jam, and could be eaten on its own. Go past Betty's Café and Tea Rooms in Harrogate, and you'll see plenty of fat rascals in the window. Well, a fat rascal kind of sums up this place.' The tone was set. 'All the colours are warm, they're super saturated. With Ros's costumes, she's trying to go with that feeling of warmth.'

Jackie says of the set, 'I feel like I live here. I love the hallway and the corridor that links all the spaces together. That was where I started, so it's a kind of journey through the house and it was the beginning of my journey.'

Mrs Hall's Domain

Into the kitchen went an Aga (which doesn't work), and a butler's sink (which does). The rooms of the studio were exaggerated by about a tenth of the original size, in order to make space for the crew as they filmed.

In Ben's scripts, Mrs Hall becomes the beating heart of Skeldale House. Jackie adapted this in her own way. 'We created a world for her in the scullery where she has a separate area that she uses to relax and write letters. That space is full of detail, props that relate to her past – the fact that she has a son, and that she has a life that happened before she arrived at Skeldale.

'Even at the start of a scene she might be tackling a project that she can then put to one side while she delivers her dialogue. For the second series we added a pantry to the set to give Mrs Hall another working space to pickle and make preserves.'

Samuel West and his Props

Jackie and Sam spoke about the smaller props for Siegfried. It could be a pipe that he holds, a penknife or a notebook, or a small book that he carries in his pocket. 'We help wherever we can and offer things that maybe he's not thought of. He does appreciate that. Again, these props give depth and context to the character.

'Siegfried served in the First World War so there is war memorabilia in his office. In the story he's lost his wife four years earlier, and there are photographs of her around the room. These props may not always be noticeable, but I really think that they help any actor's performance – they give them that world to work in.' There is also his wife's music room: 'Siegfried has made his private space. When he's in that room he gets to reflect on the past and do his bits of chaotic admin in there.'

Melissa Gallant remembers Jackie talking about the office. 'Siegfried's office was not a feature in the books,' Melissa says. 'This was Jackie and Brian's rather brilliant invention. The office sits at the centre of the set, with Siegfried's busy desk strewn with paperwork, the externalized chaos of his mind, and myriad items that represent his world and his interests. The set design was representative of Siegfried's role in the house; his domain sits in the very centre of the world of Skeldale, just as his presence, authority and somewhat chaotic tendencies sit at the centre of the show, setting the temperature in the house, creating story and influencing everything and everyone around him.'

Designing Broughton Hall for Tricki Woo

Broughton Hall, home to Mrs Pumphrey the adoring owner of Pekingese Tricki Woo, was already filled with a collection of antiques, but Jackie picked up on Mrs Pumphrey's days in India and that she was 'an eccentric in the grandest way possible'. Now, when it comes to filming, the house is adorned by Pekingese-related props, rugs and ceramic parrots.

Dame Diana Rigg liked to be given a small selection of things, such as morsels of meat that she would take out of the bowl for Tricki Woo to enjoy. 'She chose what she thought would be best for her character. And all those little touches can really help any actor, no matter how experienced, get into their role.'

Jackie says of Diana, who died in September 2020, 'She was an absolute delight to work with.' With Patricia Hodge taking over the role of Mrs Pumphrey from series two, Tricki's storyline continues to develop with humour and pathos in equal measure.

Heston Grange and The Ritz Cinema

In series two, Jackie and her team were asked to build a set for Heston Grange, home to the Alderson family. 'Several scenes were set in the kitchen and the stairwell,' she says. 'In the design for this set, I wanted to show that in contrast, Skeldale House was more up to date in terms of its décor and equipment, as Heston Grange is a remote farm without power.

'In the story Mrs Alderson passed away some years previously and all attention has been on the working of the farm since then. It leaves little time for niceties and thought for decoration. This leads to the kitchen having paint-washed plaster walls and old Georgian cupboards with a blackened range, and flagstone floors. The colour palette for the set featured the colours of the earth. The scenic painters did a lovely job, building bulgy, plastered walls and layering up thin washes of colour to give the room a real sense of age and permanence.'

They built other sets on location. 'We transformed a corner of Thirsk, taking it back to the thirties, creating a clothes shop and a motor sales shop. We used archive references to make the Ritz cinema look as it did in 1938, recreating the original sign and ticket office.' Jackie adds, 'The locals seemed delighted.'

CHAPTER EIGHT

Finding Darrowby

Brian and Richard – lead director and producer – needed to find a real-life Darrowby, home to the veterinary practice. Alf Wight based the fictional town on Thirsk, Leyburn and Richmond.

How on earth do you find a village in 2019 that can be transformed into a town in 1937? The assignment was given to Gary Barnes, the location manager for *All Creatures Great and Small*. Gary, who lives on the outskirts of Leeds, read the scripts so that he had a feel for what was required. Then he set off, driving into the Dales, and visiting all the towns and villages in Wharfedale. Along the way he took photographs to show Brian.

'The brief was to find a pretty village, but not picture-box pretty and twee,' says Gary. 'In the BBC series, Darrowby was further north.' The location of their version of Skeldale House was in Main Street, Askrigg. 'But our Darrowby had to be in the south part of the Dales, closer to Leeds.'

There were crucial factors to consider before filming. Questions had to be asked about whether a location is viable as Darrowby. Is it possible, for instance, to have control of the roads? How noisy is the area (there are beautiful villages in the Dales, but some are close to A roads used by trucks and coaches)? Is it easy for the film-makers – their vehicles filled with equipment, others carrying animals – to access the area ... and then find somewhere to park? Can the village be transformed, taken back in time to the thirties? If so, what will it cost to transform shop fronts so that they match the period, and install street signs? Will the local residents and businesses give a warm welcome?

And don't forget, All Creatures also involves the challenge of animal welfare; the second series was made during the lambing season, the busiest time of the year for farmers. 'What's more, we're always working to a budget,' says Gary. 'Someone else might stand in the middle of a village and think it's perfect, but I can look at it and see many reasons why it won't work for filming.'

With its stone-built cottages and cobbled market square, Grassington was one of those villages that Gary photographed. 'It's unique in size – smaller than other villages but it also has a commercial community. Straight away we felt that it seemed right. It really does feel like Thirsk must have felt in the thirties.'

Richard says, 'Grassington has the most beautiful square tucked away and off the main road, and so it's quieter than other villages, and is square on all sides so you can shoot all the areas of it. It is its own contained little world, but it's very, very, very beautiful.'

Grassington would be their Darrowby. 'Then,' says Richard, 'we identified a property we thought would make a good exterior for Skeldale House, the veterinary surgery, and where the main characters live. So through our fantastic location team we started approaching individual shops and homeowners within the village. Plus also getting in touch with the Chamber of Commerce for Grassington to talk through what we wanted to do. In the first instance, we would be going there for a few days filming, but obviously we're making a number of episodes sequentially with the promise of the show returning in the future.

'So we wanted to set up a relationship that gave us what we wanted, to create that fictional 1937 world of Darrowby, but that also worked with the local homeowners and the local businesses for filming.'

They removed modern street furniture, such as bollards, put up old signs and redressed the fronts of Grassington's shops. 'They have been fantastically welcoming and allowed us to use their beautiful village to create an area that's very important to us in *All Creatures Great and Small*, with the exterior of Skeldale house and the slightly wider village area as Siegfried and James, and Tristan and Mrs Hall, go about their business, perhaps go to The Drover's pub, or to the market, or even to the Darrowby Show in the village square.

'The villagers fell in love with it,' he adds. 'Some of them even asked if we could leave it that way, so that they could stay in the thirties.'

Rachel Shenton recalls going to Grassington just before filming began on the first series. 'We couldn't have asked for a better reception. When we first started filming I hadn't watched the BBC series – a bit before my time. But my mum was very excited that I was going to be in this adaptation. I didn't realize the popularity but when we were filming in Grassington the villagers were so excited that we were making All Creatures but there was a caveat: this had better be right. We felt a collective pressure.'

CHAPTER NINE

Every Shot Tells a Story

After a four-hour drive from his home in Bristol, Andy Hay arrived in Skipton and checked into a hotel beside the canal. It was October 2019 and shooting on All Creatures had started a few weeks earlier. This pretty, ancient market town on the southern edge of the Dales was by now the well-established base for the show's cast and crew. Andy's objective, the purpose of this initial trip to Yorkshire, was to prepare and plan for the shows that he'd be directing in this first series – that is, episodes four and six, as well as the Christmas special.

The coming days would be busy and exciting. Tomorrow Andy would drive deep into the countryside with the location manager, Gary Barnes. They'd check out farms, properties and breath-taking beauty spots that might make ideal settings for particular scenes. Andy would have meetings with Jackie Smith, the production designer who had created the set of Skeldale House, and Ros Little, the costume designer. At some point soon he'd meet with Andy Barrett, the veterinary consultant, and Jill and Dean Clark, the animal handlers, as well as Mark Atkinson, the horse master. There'd be get-togethers and rehearsals with the ensemble.

The night before throwing himself into the prep for *All Creatures Great and Small*, Andy relaxed in his hotel room with Herriot's book of the same name. It was a new-ish copy as Herriot was new to Andy, and he removed the bookmark and picked up the story where he'd recently left off. 'It was then that I read a particular passage about James Herriot. He visits an old man who has a dog that's also very old. To spare this much-loved canine companion any distress or suffering, James puts the dog to sleep. The manner in which he tenderly cares for the dog, and the deeply moving way the old man says goodbye to his old friend … It was all so beautifully described by the author – and in such vivid detail – that I was transported to the tiny, dimly-lit backstreet cottage. I felt like I was sitting in the room with the three of them. It was written with such utter heart that as I read the tears rolled down my face.'

For Andy, Alf Wight's writing (in the pen name of James Herriot) is a magical concoction of bits 'that make me chuckle, make me feel good about life and the world'. That evening, when he was moved to tears, he felt genuinely privileged to be involved in this television adaptation. 'I got to the end of the chapter, put down the book and thought, I'm so bloody lucky to be making this show.'

Lucky perhaps, but ahead there were enormous challenges, such as the first Christmas special that featured the birth of the Border collie puppies (see pages 38-9). In the second series, when it came to shooting the episode about Mrs Pumphrey's annual cricket match, Andy had to cope with more than his fair share of sticky wickets. Then there was episode six of the second series, which featured a storyline that required significant prep, skill and imagination …

There's trouble at the Aldersons' farm. One of the horses, Candy, is in foal and is in agony. Candy is also special because she had belonged to Helen and Jenny's late mother. On examination, James discovers that Candy's uterus is twisted. The horse has to be rolled on her back by Siegfried and Mr Alderson, from one side to the other, while James – with his hands inside the birth canal – holds the foal by its hooves (that way, he untwists the uterus, enabling the birth). It's an uncommon and life-threatening procedure for the mare and the foal, but this is an emergency and the only option for James.

'There was an unusual aspect,' says Andy. 'Mark Atkinson, the horse master, said he had a horse that could roll 180 degrees, from one side to the other, legs up in the air. But the horse was Aramis, who is a male.' Enter Pauline Fowler, the prosthetics expert, who was assigned to create a mare's rear end – complete with twitching tail to trick the viewer's eye and add believability. Says Andy, 'We had to do to some judicious editing afterwards to remove any male parts. Apologies Aramis!'

They hoped to film the scene in a barn but this presented two problems: first, horses prefer to give birth outdoors, and they don't like an audience during labour; second, there was too little space in the barn for Aramis to roll on his back from side to side. So they decided to shoot the scene in a field.

'Then I had to find a way of showing that a foal had just been born without having a new-born foal actually on set, or a prosthetic one.

A real newborn, of course, was a no-no, and a prosthetic might just look lifeless. But it was in the script and it was my job to find a way to make it a reality on screen.' They located two mares – one in Yorkshire and another in Devon – that were similar in colour to Aramis and soon to give birth. Camera operators were on standby to film these births in a similar setting to Candy's delivery; in a field with trees and dry-stone walls.

'On the day, Aramis was brilliant – he rolled four times for the camera, which meant we managed to get all the shots we needed.' Andy also wanted to shoot a natural reaction from the actors as they witnessed the foal standing minutes after 'the birth'. So he placed his laptop on a chair in the field where they were shooting the scene and played the footage of the foal's first steps which a week earlier had been captured by the camera operator on stand-by. The actors – playing Helen, Jenny and Mr Alderson, Siegfried, Tristan and James – watched the footage of the new-born foal stumbling and standing up for the first time and taking its first suckle. 'As they watched, we filmed their reactions. That meant on screen we had a connection to the birth of Candy's new foal.' Genius!

There was, however, another issue. Aramis had white fetlocks but the real mother had black fetlocks and a longer mane. Again, this called for more judicious effects in the edit suite. Only one eagle-eyed viewer wrote to point out that he'd spotted the horsey difference.

Andy sees these challenges and obstacles as an essential part of creativity. With vast experience in theatre – for a decade he was artistic director of Bristol Old Vic – he has frequently had to resolve the potential problems of performance. At times Andy, a former actor, has even stood in for actors who are unwell. 'I'll happily go on stage in their place,' he says. 'I've played all sorts to fill in for actors.'

Years ago, at Bolton's Octagon Theatre, he was directing *Jack and the Beanstalk* when he did just this – not wanting to disappoint the audience, he stood in for the actress playing the part of the giant's magic hen who was too unwell to go on. 'She was a brilliant young actress, Billie Williams [these days she's a first assistant director in TV]. For this role, she'd put on an egg-shaped papier-mâché costume and lay golden eggs from the back of it by pulling a wire in one of her wings, where her hands were hidden. I say *golden eggs* but they were made of polystyrene and bounced when they hit the stage. The trouble is, she was five foot one, and I'm five foot ten. I only just managed to squeeze into the egg-shaped costume and, being that much taller than Billie, my dignity was barely covered where the papier-mâché body ended and the yellow tights began. I couldn't properly fit my arms in the wings in order to pull the wire to lay the eggs. I could only get my elbows in, so I had to put an egg under my arm and then lay it from my armpit …'

The audience was in gales of laughter. 'After four shows of being this very odd-looking hen, the giant started playing the audience, saying things like, "That's a funny-looking hen. Has anyone ever seen a hen lay an egg from under her arm?" So we kind of made the most of it in the end.'

In theatre he enjoys the technical rehearsals, that bit when it all comes together – 'the white-hot fusion', as he says, of the actors, the music, the costumes, the set. With television, 'Sometimes – not often – you have to rehearse quickly and shoot quickly, and really that's when you rely on the good will, camaraderie and professionalism of the actors. When we get on that farm or that set and we're all there in one place – cameras, costumes, lighting, the lot – we have to marry it all together, come rain, sun or snow. That's what I love,' says Andy, whose television credits include *The Last Kingdom* and *Lucky Man*.

Preparation is crucial. 'When we begin a filming day, I want to have a very clear idea of what the day entails and how it'll be approached and executed. It's important to keep a collective momentum on set. Momentum is a creative energy and helps keeps morale high. Lose momentum and everything just dribbles away.' (Rain, meanwhile, is 'terrible because everybody vanishes into their hoods, and they tend to go into themselves'.)

He is never without a camera plan, even if it's a plan that changes on the day. The way he sees the business of a filming schedule, 'It's about eliminating the trip wires. There will always be something – "Oh my God, that gate's locked", or, "The car won't start", or a particular favourite: "That wasn't there when we reccied this location". The director is the guide, and you want the cast and crew to feel confident that you know what you're doing and, more importantly, confident they know what they're doing.'

With all of this in mind, come now to a cricket field near Harrogate, North Yorkshire. It is the summer of 2021, and the sixteen-week shoot for series two of All Creatures is drawing to an end. In these final five days Andy will be shooting the fifth episode: Mrs Pumphrey's annual cricket match (and entitled 'The Last Man In'). After weeks of poor weather, Andy, his crew and the cast are delighted that the sun is shining on them. 'Yes, the sun was a blessing,' he recalls. 'Apart from that, there were considerable challenges.'

The National Trust, for instance, had to remove a deer fence around the field (there wouldn't have been such a fence in Herriot's 1930s). This was supposed to be an event for the village of Darrowby, which meant having a decent crowd of villagers. Over the five days Andy worked with varying numbers of supporting artists (they are the artists formerly known as extras). In order to meet budgetary needs and to allow the costume and make-up teams time to prepare ahead and have everyone camera-ready when the schedule demanded it – a cast of fifteen actors plus supporting artists in period hairstyles and costumes is a mighty challenge if needed all at once! For the first and second days of filming he had twenty supporting artists; then two days of forty; for the final day eighty.

'We were shooting on a cricket field so everyone was dispersed, as you can imagine.' However, as the viewer you want to see all the characters and the crowd at the same time. This led to complications. 'Although we are used to filming shots out of order on television, in this instance it was like throwing sticks into the air and then waiting for them to drop, not knowing in which order or when they'd land. The logistics for the camera crew, the grips, moving equipment around – it was pretty crazy. Everything was filmed almost completely back to front.'

He shot the closer and mid shots when he had crowds of twenty and forty. The wide shots, which showed the whole field, were filmed on day five when Andy had that larger crowd of eighty. Meanwhile all departments, understandably, needed to know who was coming in and when. Plus, it was crucial to know precisely when the food for the cricket tea should be made and put in the marquee – 'the sandwiches and cakes wouldn't last all week'.

The challenge intensified when Samuel West injured himself … twice. (see page 125). First, Sam was practising in the nets when he pulled a muscle in his right calf. His leg had to be strapped up and he was limping. Then, and this was on the final day of the shoot, Sam began a dash from the crease of the pitch when suddenly – ouch! – the Achilles tendon in his left leg snapped. Minutes later, the show's Siegfried was on his way to Accident and Emergency at Ripon Community Hospital.

Apart from the considerable pain for poor Sam, the incident was certainly no fun for Andy and minus his Siegfried, he was left with a problem. In the story Siegfried was supposed to be run out. If Sam couldn't run for the cameras then this part of the storyline would have to change. 'I had to think quickly – *Siegfried can't just disappear in this episode. I'm going to have to rethink how I tell the story …'* In cricket terminology, Andy needed Siegfried to be out (as the batsman) so that James could be in.

'In the middle of the cricket pitch I reviewed all the shots we had so far. There was one take of Siegfried batting where he hit the ball high enough to look like it could be caught. While Sam was at the hospital we shot the reverse of that, so we had a fielder catching the ball, which appeared to be the one Siegfried had just hit. And I found another shot from a different scene of Siegfried looking dismayed, slightly disappointed, at how James was playing as a fielder. In another context it looked exactly like Siegfried was thinking, *OK, fair enough, I'm out.'*

He had solved that problem but there was a final scene, which was to have been Siegfried in the marquee, moving among the crowd. 'Remember,' says Andy, 'I had my eighty people and I thought, I'm going to bloody use them whatever happens.'

Valiantly Sam returned to the shoot with a full plaster on his leg and, at this point, the rickshaw came in handy. The rickshaw is a two-wheeled piece of camera equipment that can be pushed at speed to keep up with running actors. Now, with injured Sam back at the cricket field, the rickshaw acquired a new purpose. It became a sort of sedan chair (on wheels). 'We placed Sam on the rickshaw and wheeled him to the shade of the marquee and kept him there until we were ready to shoot the final scene.' Andy had revised his ideas of how to shoot

Left: Alf Wight, the real James Herriot, in his beloved Yorkshire.

Below: The Yorkshire village of Yockenthwaite becomes the Alderson farm in the late 1930s.

Above: Alf Wight with his much-loved dog, Bodie. Below left: Nicholas Ralph as James Herriot on the set of *All Creatures Great and Small*. Below right: Alf Wight at work.

Above: Nicholas Ralph as James Herriot on set with lead director Brian Percival.
Below: The cobbled streets of Grassington provide the perfect outdoor backdrop
to Skeldale House.

Above left: Anna Madeley as Mrs Hall. Above right: Production Designer Jackie Smith's plans for creating the set of Skeldale House. Below left: Samuel West as Siegfried Farnon. Below right: The real-life inspiration for the dispensary at Skeldale House.

Above: Alf Wight, centre, with the inspiration for the Farnon brothers, Siegfried (Donald Sinclair, left) and Tristan (Brian Sinclair, right). Below: The first meeting between Jim Wight (bottom right), his sister Rosie Page (top middle) and members of the cast and crew.

Above: Mrs Pumphrey was played by the late Dame Diana Rigg in the first series; in series two (below) Patricia Hodge took on the role of the redoubtable guardian of Tricki Woo (played by Derek).

Above: Anna Madeley (as Mrs Hall) with Frank. Right: Imogen Clawson (as Jenny Alderson) with 'Scruff'. Bottom: Callum Woodhouse (as Tristan Farnon) with 'Jess'.

Above: Matthew Lewis and Nicholas Ralph with Highlee Percy Pickle (or 'Monty' as we know him). Right: Rachel Shenton with Aramis (aka 'Candy').

the last scene: he'd place injured Sam on a tall bar stool in front of the small wooden cricket pavilion, and shoot him waist up through an open shutter from inside the pavilion. 'First, I needed to check with Sam if this was going to work for him in the full cast. I said, "Sam, I've worked out how I want to shoot the last scene. Can you sit on a bar stool and drink a pint of bitter?"'

From his rickshaw recliner, Sam said, 'Yes Andy, I can do that.'

'So that's what we did, and Sam didn't have to move at all. Instead everything moved around him. It worked.'

He pauses. Then adds, 'We can all smile about it now.'

'Every shot should tell a story,' says Andy. 'What is the audience seeing in the frame? Perhaps it's a character's emotional journey you want to bring out, or some action or the setting in the background.' He searches constantly to show detail within the shot. 'It might be in the way one sentence is said by an actor, or it might be a pause or a look. Or it could mean shooting from behind, or from a particular angle, to capture what the actor or the mood of the scene is looking to convey, and therefore making it a more telling experience for the viewer.

He feels 'a responsibility to the viewer, to the audience. They've given up their time to see what we've created. So I am constantly thinking of the audience.' Andy adds, 'I don't want them to say, "That was all right." I want them to say, "That was really good."'

PART THREE
The Cast

'It's times like this which remind me how grateful I am for everything I have … Not the practice, or the house or the beautiful countryside, or any other thing. It's the people. Infuriating as you all are, I'm rather fond of you. And well, there's that – so, well … Merry bloody Christmas.'

– SIEGFRIED FARNON

CHAPTER TEN

The Wonderful
Tricki Woo (played by Derek)

'You're never going to win a scene when Derek's in it.'

– NICHOLAS RALPH

Mrs Pumphrey and her ridiculously spoilt Pekingese Tricki Woo add plenty of humour to *All Creatures Great and Small*. James Herriot creator Alf Wight based them mostly on Miss Marjorie Warner and her Pekingese, called Bambi. Jim Wight says, 'Many of Dad's characters were a composite of people he came across. Mrs Pumphrey was probably 80 per cent Miss Warner. She was a very pleasant woman, terribly fussy. She'd say, "Bambi is so tired of chicken." And this was in the days when chicken was a luxury, like smoked salmon is today. She realized that Mrs Pumphrey was her and was very nice about it.' Miss Warner and Bambi lived in the village of Sowerby, a couple of miles from the Sinclair-Wight practice in Thirsk.

Tricki Stories

Although Marjorie Warner was a spinster, Herriot presents Mrs Pumphrey as the widow of a wealthy beer baron who lives in a beautiful house, with servants, on the outskirts of Darrowby (in the series, Broughton Hall in Skipton is used as the mansion's film set).

In the first series of All Creatures, we see that the little dog is indulged with Egyptian cotton sheets and edible delicacies that include chicken, plum duff pudding and – his favourite – trifle. James becomes the Peke's 'Uncle Herriot', and receives hampers, kippers and grapes from Tricki.

The show's series bible, written by Ben Vanstone, sets out a storyline for episode two of the first series: 'James treats Mrs Pumphrey's dog Tricki Woo and is rewarded with an invitation to her party. James meets Mrs Pumphrey's dog Tricki Woo and admonishes her for overfeeding the little Pekingese. Mrs Pumphrey and Tricki Woo take an instant liking to James and are eager to get to know Darrowby's newest vet …'

The acting star who plays Tricki Woo

Tricki is played by Derek, a Pekingese who lives with Jill Clark in Lincolnshire. She and her son Dean run 1st Choice Animals, which supplies the animal stars of All Creatures. Derek features on the credits, along with the main cast members.

Jill Clark on meeting Derek

'He walked in and I knew he was right the minute I saw him. I was looking for a Tricki Woo, and Pekes from rescue homes are very hard to come by, but I didn't give up. Eventually, a lady turned up at my house. At the time, I was in the garden and in the middle of a barbecue with friends. She was carrying Derek in her arms.

'The lady said, "My sister and I have been breeding Pekes for years and years and years. She had the dogs and I had the bitches." Then she added, "But now my sister has died, so we've got to re-home some of the dogs." My friends and I sat there as the lady plonked Derek on the table. Then she pulled out a bottle of talcum powder and smothered him in it. "This is how you look after them," she said, as she showered him in the stuff. It flew all over the barbecue. We were astonished, trying to take it all in. "He's going to be very, very happy here," she said. "I'm going to leave him with you."

'And with that, she turned and left. Off she went. And that was how I came to meet Derek. He'd arrived.'

Jill Clark on Derek's acting talents

'What can I say! Derek is just a natural. I knew he was just right for the part of Tricki Woo. He has not a care in the world. All he needs is to be lifted into place and then he'll get on with it himself. Everybody adores him because he's such a nice, laid-back, easy-going dog and, just like Tricki Woo, Derek is very, very spoilt. He has a big hairdryer – it's like a leaf blower – which I use after he's had a bath. It gets out the undercoat and dries him very quickly. He also has his own little leather sofa. But that's how it should be – he's such a sweetheart.'

Callum Woodhouse on Derek

'There are plenty of laughs all day when you're working with Derek. He's so well trained and he's hilarious, just the silliest dog I've ever seen. He's also like number one on the call sheet – higher than all of us. Derek is the most important member of the cast!'

Nicholas Ralph on Derek

'Derek's incredible to work with … and he gets better all the time. There's a scene in series two when I have to anaesthetize him. The director calls for a hush before the cameras roll: "Quiet, quiet." And then there I am with Derek, pretending to do the injection. Suddenly, I could see Derek's eyelids slowly dropping, dropping, dropping … *And he went to sleep.* It was perfect timing. And then the director said, "Cut!" Derek shot up, wide awake. He was like, "What's up?" It was an astonishing piece of acting. He's brilliant and takes all the pressure off the other actors. You're never going to win a scene when Derek is in it.'

For more on Tricki Woo, read Patricia Hodge's account (page 141) of playing Mrs Pumphrey.

CHAPTER ELEVEN

Anna Madeley on Mrs Hall and Skeldale House

Skeldale House, it was felt, was imbalanced, and certainly for a modern-day audience. There were the three men – Siegfried Farnon and his younger brother, Tristan, as well as James. In this adaptation, it was decided that Mrs Hall would be a fully formed character with her own story. First one up in the morning and the last to bed, sharp as a tack and with a wry sense of humour, she became the unsung hero of Skeldale House, the heart of the family. Anna Madeley gives a unique insight into life as Mrs Hall.

From the very start, I was fascinated by Mrs Hall. She just leapt off the page for me. Emotionally very brave, she is a fully rounded human being … and with vulnerabilities. No, she doesn't always get it right. Yes, sometimes she says the wrong thing. But Mrs Hall always means well and doesn't set out to stamp her values all over everybody else. She is a more nuanced, caring character, seeking out the goodness in others. She's no ordinary housekeeper.

We know that she had her time in the Wrens, married and then left her husband. She has a grown-up son, Edward, although he is elusive. So her life had fallen apart and what we see now is a woman who has come out the other side of an ordeal. There is sadness in terms of her relationship with Edward but, fundamentally, Mrs Hall is quite happy, and enjoys the role of housekeeper.

She has found a good place to be, is valued, has fun and she has the community around her. She is a character who is aware of her own identity, her self-worth, and I think there is something appealing about her degree of independence. And this spurred something in series two – whether she might entertain meeting someone new, contemplate the possibilities of opening up her heart again, live a full life.

True, the men of the house go out in all weathers, encountering all sorts of obstacles, but back at home Mrs Hall has a full-time job, and it's physically demanding, without the high-tech machinery and gadgets that we have today. The washing is an endless, heavy-duty job with the mangle. She cooks and bakes, and is out shopping on a daily basis. Life for Mrs Hall is very much in the present.

And as I play her going about her daily work, I have to keep in mind that she is an employee of Mr Farnon. That is her security. It is something to be protected and we always have to go back to that point and ask ourselves: How often can she speak her mind? Is it worth her sticking her neck out and taking that risk, knowing that ultimately it's for the right reason. There's a fine line for her to tread.

Likewise with Tristan and James, she is not their mother although she has quite a maternal role. This aspect of her character develops in series two, again with the questions being asked: How much can Mrs Hall be involved? How much can she steer them? And at what point does she draw the line?'

As part of Ben Vanstone's fantastic adaptation to the surrogate family of Skeldale House, he raised Mr Farnon's age and lowered Mrs Hall's age (in his books, James Herriot guesses that she's about sixty). This creates a more parental set-up between them, adding a fruitful dynamic to this multi-generational show. With Mrs Hall as the maternal figure at the centre of the house, it opens up the inner, emotional life of the characters. The books are from James's perspective but, in the series, that 'family' dynamic is an excellent way of allowing the audience in.

James Herriot gave that wonderful description of Mrs Hall at the beginning of All Creatures – a figure of grim benevolence who welcomes James in, surrounded by her servile pack of dogs, and then disappears down the hallway into the back of the house. That is as much as you get of her. She is firmly in the scullery, barely seen and rarely heard. For this adaptation, she has a strong back-story – and, of course, she has to appeal to the modern-day audience. The joy of an ongoing character is in that evolution; there are still things to reveal about Mrs Hall. I was cast very late. Two weeks before I was on set, in fact. I auditioned by sending a self-tape that I made at home with Tom Godwin, my lovely actor friend. Tom explained cricket to me, operated the camera, and played Siegfried, Tristan, and James off camera for me. That was my audition – I didn't actually meet anyone! Thankfully, however, I'd previously worked with Brian Percival (on *The Old Curiosity Shop*). So once I was cast I had a productive phone conversation with Brian, in which we discussed the character and the piece as a whole. I also had a long chat with Melissa Gallant and Richard Burrell, who very warmly welcomed me into the All Creatures' family.

In terms of preparing for the role, it was partly that instinctive response to the script and partly about understanding who she is, not just her story but also, for instance, her wardrobe. As housekeeper she is opening the door to visitors and she's also the person who scrubs the floors. We had to find a balance for her outfits. We were lucky – I had one day's shoot, for the scene in episode one in which James has returned from the pub, a little worse for wear, and is with the cats at the back of the house. The other cast members had met, done a read-through, visited the farms and done a vet 'boot-camp'. I was thrown straight in, wielding a cricket bat in the middle of the night at a delightfully inebriated James Herriot. Ros Little, the fabulous costume designer, got my dressing gown and Wellington boots ready for that first day, and – an added bonus – the scene was set in the middle of the night so fortunately Mrs Hall was able to have her hair pinned up for bed.

Then I wasn't needed on the shoot for another two weeks. This fortnight's grace allowed us to consider what Mrs Hall's hair should be like, and there were lengthy debates about the rest of her wardrobe. Often I'd try on a costume and say, 'I look like my grandma.'

It was all about finding ways in which she can present herself, but on a housekeeper's wages so no vast wardrobe. It's more about pinny on, pinny off. And I like the fact we found that lovely, old cardigan. Once you establish a sort of palette of clothes, it becomes quite easy. You try on other things and think, 'Does this fit in and does it suit me?' We stick to that strong and important sense of 'homemade' and Ros has the knowledge and experience to achieve that.

During that two-week period of trying on clothes, there was another issue. Mrs Hall's hair. Lots of wigs had been made for me, courtesy of Lisa Parkinson and her hair and make-up team. I tried them all. Umpteen photographs of me in wigs were sent to the producers. Eventually, we found the solution. My own hair and that's what you see on the show – no wig for me!

Then the accent. I worked on that with the brilliant Natalie Grady, who's a voice and accent coach. I'd met her before, when we worked on the BBC drama, *Time*. Natalie is fantastic at helping me find my way. We go through the script and think about alternative ways to approach the dialect so that I find my own nuance. Natalie is always around if I need her, and we're also surrounded by the Yorkshire crew, which is helpful because I'm constantly hearing their voices. Sometimes finding the accent is simply to do with the phrasing, or how a word may be clipped. These tiny details are so useful, and as you can imagine, it's an ongoing part of the job.

The sense of family that we see in All Creatures is echoed off-set and I feel that I've lucked out massively. At the heart of it are these stories that everybody enjoys and loves – they're precious to us and perhaps a motivation for getting on so well. We certainly have a good laugh together and for series one we went out for dinner a lot, though that wasn't really possible when it came to series two and Covid restrictions. It was dinner for one at home for quite a while, although as the season went on we managed a few cosy outdoor dinners under blankets.

Is Mrs Hall similar to me? I like her sense of fun, though she's probably more grounded than me. She's often seen knitting and sewing at her little machine in the corner. I'm pleased I did some sewing when I was younger so I had a head start! As a child I went on activity holidays in Somerset, where I learnt kayaking, archery and how to shoot a rifle. I was surprised to discover that these activities were good preparation for playing Mrs Hall. Rifle shooting at the summer fair! You never know what skills might come in handy. Like Mrs Hall, I am adventurous and happy to try something new. Then there's the cooking ... Mrs Hall prepares hearty roast dinners and elaborate puddings. These are made by Bethany Heald, the show's food stylist and home economist. She prepares feasts for Skeldale House in a pretty makeshift kitchen. When you see Tristan tucking into his roast potatoes, it's no problem at all because the food is just delicious. I never complain about eating breakfast for one take after another, though you do have to think ahead and be careful not to get too full.

Bethany is brilliant and has load of tricks. In series two Mrs Hall makes blancmange – Tristan's favourite – and Bethany taught me to pipe the cream on it. She put extra gelatine in the mix so that it was almost solid and impossible to wobble and split.

During the shoot I stay in a rented house and always intend to make it more like a home. That doesn't happen. The days are so busy I barely see the place. However, my children gave me a beautiful orchid and it stayed with me, in the house. I managed to keep it alive for the whole period of filming series two, and then it came home with me. I'm not particularly green-fingered and plants tend to die on me, but this orchid is alive, thriving and extremely content.

CHAPTER TWELVE

The Audition with Nicholas Ralph

Intrigued, enthralled, slightly nervous. And then
immense respect for the creator of Herriot – Nicholas
Ralph reflects on the sequence of events that led to
him becoming the world's most famous vet.

'Would you like to audition for the role of James
Herriot?' It was an email request from my agent,
Phoebe Trousdell, in the spring of 2019.

You need a picture of what was happening in my life at that time.
I was in a dressing room at The National Theatre of Scotland. I had
landed the lead in a new play, *Interference*, and it was just a few
days before the curtain would go up, the show would go on. And

then, as we prepared for rehearsals, that email about auditioning for a new TV series came through. James who? I hadn't come across the books of James Herriot, or even heard of the man or his creator, Alf Wight.

And I did not know, when I got that call, that I was in a theatre just six miles from the homes where Alf grew up – first at 2172 Dumbarton Road before moving to 724 Anniesland Road. Or that I was only a mile from what had been the Glasgow Veterinary College, in Buccleuch Street, where Alf studied to become a vet. Before he moved to Yorkshire, Glasgow was the place he called home, although he was born in Sunderland.

Nor did I know that my mother and Aunt Moira were huge fans of Herriot. Somehow that had passed me by. I was also blissfully unaware that Moira's husband, my Uncle Henry, had the entire collection of Herriot's books. And, when I took Phoebe's call, I hadn't a clue that, as a lad, Henry had gone to Thirsk, visited that veterinary practice at 23 Kirkgate *and* met Alf, the man behind James Herriot. Alf had kindly signed the books for young Henry and this cherished collection has remained my uncle's pride and joy, and rightly so. How many times must I have walked past that signed collection at his home?

Though I was still unaware of Herriot and Alf, and that my uncle had met him. All of this I had to discover a while after Phoebe's email.

I said to Phoebe, 'Really? I don't have a lot of time. It's Friday now. We're opening Tuesday night. But I will do my best!'

'Just relax, you're really right for it,' she insisted. 'Go for it.'

'Ok, yes, sure,' I said. It was one of those moments when you question what you have got yourself into. After the exchange of emails I considered the dilemma – do I focus on the play or prepare for the audition? The play demanded so much of my time but any spare time I had I was preparing for the audition scenes. I did a bit of research into Herriot and All Creatures but for the time being I couldn't do much more than a bit of googling. This is going to be quite tricky, I was thinking. How am I going to make these balance? Do I really have time to do this audition justice? It was a strange position to be in.

As I was unaware of James Herriot, I didn't fully appreciate the sheer size and scale of the role for which I was auditioning. Then eleven pages of script were sent to me so I could learn my lines for the audition, and I realized there was a lot to learn. I read the script … and that was it. I was struck by the quality of the writing. It was an incredible script, easily the best writing that I'd come across for a long time.

As I read, I felt an immediate connection to the character of James Herriot. I knew who this person was and could identify with him. The sides that I'd been sent included the scene in the first episode, in which James is on the farm with Helen and, awkwardly, he scrambles up a drystone wall to get away from Clive, the bull. James ends up on the wall holding a chicken and Helen says, 'Need a hand down?' He hops down from the wall, embarrassed. I also admired his strength. He had such compassion and patience but he also had a backbone and would stand up to anyone, especially when it came to the wellbeing of an animal. I really respected and loved that about the character of James. Little moments like that provided the connection between James and me. I could empathize with him. The scripts were really funny, with comical scenes with James and Siegfried. As I was reading I was laughing out loud (although there was also a poignant scene with Mrs Hall).

This is really good, I said to myself. There was a balance of light and funny, with drama, which I absolutely love, especially TV. When the comedy-drama amalgamation is right, you have the audience laughing one minute and close to tears the next. That's the stuff that I love to watch too: drama, comedy and those moments that touch you.

With the characters being based on real people, they are three-dimensional, and the relationships are so strong. I was reading the lines and the dialogue was incredible. At that moment I was fully engaged. I wanted to play James Herriot. Every spare minute that I had – even in the dressing room before dress rehearsals – I went over the script. Members of the cast in the play helped me by reading the lines of All Creatures' other characters. I desperately wanted this part.

Tuesday came. Having done as much work as I could over the last few days, I was full of gusto, fired up with enthusiasm, ready to go. The audition was overseen by the show's casting directors, Beverley Keogh and David Martin, who were lovely. I performed the pieces and was super happy with them, and then the three of us chatted for fifteen minutes or so. I came out buzzing and, in turn, that audition gave me fuel for the opening performance that evening of *Interference*.

It wasn't until after the first audition that I managed to fully immerse myself in the world of James Herriot. I found out a lot more about the man, the books and the previous series. I sat down to read *All Creatures Great and Small*, which comprised two books: *If Only They Could Talk* and *It Shouldn't Happen to a Vet*. I also got a copy of *All Things Bright and Beautiful*, but didn't want to read too far ahead, so it remained unopened on my bedside table.

Like many actors, I am superstitious. If I tell people about a big audition, I might jinx it. So I like to keep my cards close to my chest, and don't tell a soul about what might be coming up. But with this part, when I had been recalled for a second audition, I was talking to a friend and broke my golden rule. I told him about the second audition and said something like, 'I'm kind of up for this thing.' Well, he told my folks. Meanwhile, the next play I was doing – a UK tour of *549: Scots of the Spanish Civil War* – took me to Eden Court Theatre in Inverness (near my hometown of Nairn), and my parents came to see it. After the performance they said, 'Nick, you never told us about *All Creatures Great and Small*.'

I said, 'Don't ask me about it. I don't want to jinx it.' That was it; they were constantly waiting for a phone call with good news. They weren't the only ones.

After the auditions, and with the UK tour having run its course, I took a break. I went on holiday with a friend to Marseille, in the South of France, to have fun open-sea kayaking in the Mediterranean. Much of the time was spent zooming from one island to another. Anyway, one day we stopped for lunch. Better check my phone, I thought, although the signal was extremely patchy. And then – ping, ping, ping – a stream of emails came hurtling through. I saw there were missed calls and an email from Phoebe, but about a different role, a smaller part. 'Could you give me a call?' But why the urgency? I phoned.

Phoebe's opening words: 'I wanted to call you to congratulate you on getting the part of James Herriot.'

'What?!' I was shouting with joy. In fact, the crowd looked over and asked my friend, 'What's going on with him?' My friend said, 'Maybe he's got this role he was going up for.' And so, with that beautiful view of the islands in the sea around Marseille, I shared the good news.

My parents were over the moon. Mum and Dad were bouncing off the ceiling and my uncle, aunt and all the family were particularly delighted that I'd landed the part.

Next there was what is known as 'the chemistry read'. There were ten actresses, each auditioning for the part of Helen Alderson. One by one they came into the room, and I did the scenes with each Helen. I felt a real connection to one of them, and not only in the room but outside as well, in between the scenes. We got on, we clicked. She was down-to-earth and easy to talk to.

I remember she didn't have a bottle of water and we were about to do a scene and she asked, 'Do you mind if I have a sip of your water?' I said, 'Of course, you don't need to ask. Just take it when you need it.' In front of the camera it was easy too. We seemed to find that balance as James and Helen, and worked well together. At the end, I was asked if, instinctively, any of the ten actresses stood out. 'Yes,' I said, 'Rachel Shenton.'

Skip forward to the second part of the chemistry read – the final round, if you like. This time it was just Rachel and me (with a small audience that included Beverley and David, lead director Brian Percival, producer Richard Burrell and executive producer Melissa Gallant). We were on a studio set, and once again the two of us went through the scenes. Just James and Helen. Still today Rachel and

I talk about how much we enjoyed the process. It is not often like that. Afterwards everyone seemed very happy and it felt good and easy to play the scenes with Rachel.

What I would later discover is that the producers had been determined to have a James who spoke with a Scottish burr, just as Alf Wight did all his life. (Melissa Gallant said, 'Beverley and David had headed to Scotland to see if they could find that actor. Then they sent Brian, Richard and myself electronic links to about forty audition tapes, longlisted from about sixty. Brian had said to me, "We'll know him when we see him."')

Brian had also said that they would likely find an actor with similar qualities to that of the character. From the forty on the long-list, they had selected eight for further auditions. I am delighted to say that they all had the same reaction. (Melissa later explained: 'There was Brian in the room, me watching the tapes on the laptop … And there he was… James Herriot. We'd found him. Brian, in the room, knew instantly that it was Nick. And we felt the same viewing on a screen. It was an exciting moment.' She also said, 'Nick was actually not unlike Alf. Kind, extremely hard-working and good at his craft, with a twinkle in his eye.')

By then I had upped the research, thrown myself into it, eager to learn as much as I could about Alf Wight and James. I read *The Real James Herriot*, which is Jim Wight's biography of his father.

I beefed up the online research, every day discovering more and more about Herriot's world. I began to watch the BBC series of *All Creatures Great and Small*, but only saw the first episode: you don't want to watch too much in case you end up inadvertently copying what another actor has done. I only needed a flavour of that series. I watched episodes of *The Yorkshire Vet*, as well as YouTube videos of the veterinary procedures that were coming up, like birthing a calf.

There was a particularly memorable moment which came about after I'd been googling Herriot. I discovered that there was information of Alf's time at the veterinary college at the Glasgow University archive. I emailed: 'Can I come and visit?' Yes, no problem. Off I went and was mesmerized as I read through Alf 's records of his days at Glasgow Veterinary College in the 1930s. There were his report cards, showing his grades and attendance. I had learned from Jim's biography that Alf was unwell when he was at vet school. Sure enough, I saw from his report card that, of everyone who had finished the course, he had one of the highest absences for many classes.

But I was particularly interested to see that when it came to grades, he was within the top three students. Instantly, that gave me an appreciation for the hard work this man put in, as well as his intelligence and his passion for the subject. If he didn't have those qualities, he couldn't have accomplished what he did. At that moment, I was even more impressed by Alf Wight, and I said to myself, 'I really want to do a good job.'

CHAPTER THIRTEEN

Being Siegfried with Samuel West

Samuel West delivers a masterclass on acting, what goes into making a role, some tricks of the trade ... and that cricket scene which left him with more than a sticky wicket.

When I am offered a part, the first question is: Do I want to play it? With *All Creatures Great and Small*, that question was quite simple to answer. I'd enjoyed the BBC series of the seventies and eighties. And I loved the books. I also enjoyed the fact that All Creatures has, as they say, brand recognition.

In addition, I'd also loved the portrayals of Siegfried Farnon by at least two of my predecessors. Anthony Hopkins and Robert Hardy were actors I'd worked with and admired enormously. I don't feel capable of touching the hem of either's garment. So, the big question was: Could I have a go?

The initial conversations that I had with the producer Richard Burrell and lead director Brian Percival, coupled with a large A4 document by writer Ben Vanstone, convinced me very quickly that we were in good hands. I felt that in every molecule of the production. There was an amount of care and of taste which showed that they were thinking about the texture of the series and the backstory of the characters. I liked Brian's vision of it as a drama with jokes and animals, rather than a sort of soapy sitcom set in an undefined, slightly rose-tinted past, although I do think it is slightly rose-tinted and all the better for that. Ben's decisions to incorporate elements from Donald Sinclair's life, rather than going directly for the portrayal in the Herriot books, made the Siegfried character deeper, more complex, more flawed.

I liked that he was funny, or at least that he should be funny. There were clearly laughs to be got with the words, and I don't get asked to do that very often. That was a big bonus. And I realized quite quickly that he would take what I call a lot of playing. Siegfried is a man of emotional size, and that doesn't mean over the top. Robert Hardy manages to do this in almost everything he does; manages to be larger than life without ever being untruthful. I tend to approach things from below rather than above.

So the answer to the first question was, unquestionably, yes, I would like to play the role. But the timing ... When I started as Siegfried I was just reaching the end of another job, directing *The Watsons*, a play written by my darling partner, Laura Wade, and adapted from Jane Austen's unfinished novel. The play was about to preview in London when we were due to begin All Creatures.

The producers of All Creatures were very generous and kind to delay the start of filming so that I could finish rehearsals in London, and be there for the preview, and then travel up to Yorkshire to play Siegfried. Which I did, being driven through the night on a Friday in the middle of September to film for day one of the shoot, Saturday 14 September. There were scenes that they simply couldn't do any other way. Perhaps looking at those scenes again, I would rather do them again. Not because they aren't beautifully shot by Erik Molberg Hansen or beautifully directed by Brian, but because on that Saturday I was slightly feeling my way towards the characterization.

As an actor you don't often get the chance to play again a part that you really enjoy (and one which some people seem to enjoy you playing). If it's a play, the play is over. If it's a single drama, finished. Although, at this moment, I think we are at the very beginning of the things that Siegfried could embody, a lot of what he has embodied so far have been great fun to play.

When approaching a role, a lot of actors – a lot of very brilliant actors – create the character by sort of blanking themselves out and then painting something else on top. Often they are great performances. I am either too shy or not skilled enough to do that. So I tend to approach the character like this: keep what's the same with you and change what's different.

I try to make people believe that Siegfried is the man for whom the potential of the things he could do in a given situation is pretty much infinite. Early on I met Jim Wight and Rosie Page, Alf's son and daughter, and they told me stories about Donald Sinclair, the real-life inspiration for Siegfried. One of the best stories was about

Donald discharging a shotgun into the wall of the sitting room at the veterinary practice in Thirsk because he was bored with his guest – a drug rep – and wanted him to go home.

And I thought, first of all, how wonderful to have that idea and not just to think wouldn't it be nice *if I could actually to do it*. And how nice to play somebody who *does* do that. And then, even if I am not given that as a 'piece of business', do we believe that I might do it, in certain circumstances – if it was boring and there happened to be a shotgun handy? The answer, I think, is yes (given I wasn't particularly fond of the wall I was shooting at, or even if I was extremely fond of the wall).

Based on what we know of Donald, it follows that Siegfried has to be eccentric, but on the right side of mad. He also has to be played by me and, as I have said, I have to keep what's the same and change what's different.

Let's say, for instance, that you are playing Shakespeare's Richard III and he says that he was sent into the world deformed, misshapen … Then you read a biography of Richard III which says forget everything that you knew about him; he didn't have a hunchback and he was very nice. And you think, well, unfortunately I do have a hunchback and I am not particularly nice. I am playing this person and so he is both of those things. Therefore, as I have said, you keep what is useful and discard the rest.

If Donald were around, I would ask him about his dad, and add: 'What made you go into the veterinary profession?' In the second series Ben makes me refer to veterinary science as a 'calling', and I think that's absolutely right. I feel that's the way Siegfried sees it, and I think that's the way that a lot of vets see it too, perhaps more so then.

I think we found something interesting in him being widowed and being slightly blocked about his loss. Clearly he hasn't got children, which Donald didn't, and that's very important. Did he want children? Or did James become his son? Was Tristan more like his annoying child rather than a brother? Or did the animals become, in a way, his children?

You look for the opposites. So you look for the patience in a man with a short temper. You look for the deep love in a man who is possibly dismissive and gruff some of the time. I am sure that Donald had all of those things.

I would also ask him about his sadness. Jim told me about one of the last conversations that he had with Donald. The phone rang, Jim answered and he knew it was Donald because he never began with 'Hello'. He said, 'It's Donald. I'm fed up about your dad.' Not long after he was dead. Clearly there was a sadness to him. All of those things are part of him, and I have probably just tried to listen to Jim and Rosie's stories, and get a couple of things that reminded me of him.

Jim and Rosie told me that Donald was quite a hypochondriac and, at the first sign of a cold, he'd put a spotted red handkerchief around his neck and knot it. Well, I got a cold quite quickly after we started filming (later to become laryngitis, and I had to be signed off for a few days). I remembered what Jim and Rosie had said about Donald, and asked the costume department, 'Would you mind if, instead of having a tie or cravat, I wear this knotted red and white handkerchief?'

It looked quite good for the period, so I put it on. Rosie and Jim noticed. I'm pleased they knew that I'd been listening to them, and that I did something that would remind them of Donald. It was true to the man.

I had to find the voice for Siegfried. I took it largely from the way he speaks in the books, and it's an interesting question as to whether he has an accent or not. He could have a slight Scottish accent, and he could even sound slightly Yorkshire. However, I thought that as James is truly Scottish, it was more interesting for Siegfried to sound different from the Yorkshires and the Scots. So I made him RP – received pronunciation, or standard English – partly because I am comfortable with it, and partly because we thought that would be sensible.

Siegfried is quite plain-speaking at times, but Ben also gives him rhetorical flourishes. A lot of the time getting laughs with the words is about having a line that you can deliver firstly as if you understand it, and secondly so that you can phrase it correctly or lightly. And a line is often funnier if it doesn't have too many words. It's about being able to tickle an audience's ears. Jimmy Carr says that when he tells a joke he works on a punchline over and over again so it has the fewest number of words in it, to get the point across. It's just funnier.

One of the things that struck me about Siegfried was how quick he was, not necessarily intelligent but his speed of thought and his speed of moving. So I also felt that was something I could tie into.

It's been very interesting working with Maimie (McCoy, who plays Dorothy) and Anna, two very brilliant and considerable actors, one of whom makes Siegfried consider his romantic life when Dorothy says it is sad sometimes that we have to 'get back on the horse'. Meanwhile, Mrs Hall is so much more than a housekeeper. She is surpassingly wise, extremely eagle-eyed and very good at playing the game of allowing Siegfried to be the king of his castle, although he does understand that nothing really goes on in the house without her knowledge or say-so. Even though they call each other Mr Farnon and Mrs Hall, and she is strictly speaking an employee, he is very fortunate in having a trusted confidante and friend who is female – quite unusual for that period. I like the fact that these are two people who, despite their considerable social differences, speak to each other as equals without using each other's first names.

Cricket is a game that I am meant to be able to play, but about fifteen years ago I stopped playing regularly. Instead I started birdwatching. That was because Sundays can either be cricket or birding, but they don't really mix. Episode five of the second series features a cricket match in which Siegfried bowls and bats, and he is supposed to run himself out so that … I better not spoil the rest.

Well, this match for the show took me back to the game that I'd given up for birdwatching. It was 2 July 2021, a day that I'll remember for a very long time, and it happened to be the last day of sixteen weeks of filming. We went to the cricket nets at a pitch near Harrogate and, when I came to practise my run-up for the first time in many years, I pulled a muscle in my right calf. Like an idiot I hadn't warmed up properly. I had to have my right calf strapped up.

Next, I was on camera and hit a shot and had to go for a run. I set off and at that moment my right leg didn't give me enough support. Snap! The Achilles tendon in my left leg broke. It's like having your strings cut. You suddenly realize what gravity is doing for you. I needed to have a plaster cast fitted on my foot, and that was done by noon. I returned in a cast and shot the final scene from the waist up. Then, and very kindly, they drove me home in my own car and I woke up on 3 July, the start of having to spend about four weeks horizontal.

So now I was in a boot, recovering from this broken Achilles tendon, which had been operated on. I wouldn't be able to run for six months or jump for nine. I couldn't wait to be back in the gym and stretch myself, and just remind myself that I am not quite seventy-five yet.

I'd taken a few months off Twitter, which is a shame because I thought of two really good puns to finish the second series. First, a photo of my foot with the caption: 'This is our newest cast member.' Second, me holding up the final slate, with wrapped foot in view, and saying, 'It's a wrap.' Meanwhile, they didn't sack me from my next two jobs, which was kind. And no, they didn't give me the boot.

CHAPTER FOURTEEN

Rachel Shenton as Helen Alderson

Rachel Shenton explains how you get to grips with your character, why working with a two-tonne bull isn't such a problem and where the series' homely atmosphere really begins. She insists that, of all the characters, Helen has the best wardrobe. 'It's like being transported back in time,' she says, adding that Helen's voice is frequently playing in her head. 'That lovely, warm, thick Yorkshire accent … I can't help but take it on a bit.'

'**M**en moan, babies cry, women cope.' 'That's what my grandma used to say. She was from Stoke-on-Trent, in the Midlands, but it's a phrase that's resonates with

Yorkshire women like Helen Alderson. And that's what I wanted to create with Helen – a quiet strength mixed with a strong moral compass. My grandparents had a tough, working-class, get-on-with-it mentality and that's the spirit I wanted to capture.

But where did this all begin, how did I become Helen? It goes back to late spring in 2019, when lots of Helens were being auditioned and, of course, I was one of them. (They found their James before they got their Helen, so by then Nick had been cast.) At these auditions there isn't much time for a long chat; you have the material and off you go. But even at the reading with Nick, I actually felt very relaxed as if he was more someone that I might have known instead of a complete stranger.

When we next met, it was just the two of us having a read-through, the 'chemistry read'. Often these can be awkward because you feel under pressure to create a chemistry. So as actors you try to make the scenes look lovely and authentic, as if you really care about each other very deeply. Of course the reality is that you are two strangers. But with Nick there was definitely a rapport, and that easiness I'd felt when we first met. We were comfortable together, exchanging a bit of light banter as if we knew each other well.

I was excited about working with Nick, and it has been really wonderful discovering our characters together as we embarked on this journey. The really lovely thing is that when I met the other main cast members – Sam, Callum and Anna – I felt exactly as I had when I first met Nick. These were all people I felt immediately comfortable with, and we became really close very quickly. This definitely feeds onto the screen and creates its own magic. Viewers say, 'It looks like you guys get on really well.' And it's true. We care about the characters and we care about each other. During the making of series one, we'd all go out a couple of times a week for our 'family dinners' at The Woolly Sheep in Skipton.

We are a community on- and off-screen, and it's clear to the viewer. Usually I quite like a day off work, but this is the only job where I look at the schedule, see that I have a day off and think, 'Actually, I'd prefer it if I were in.' And when I'm not working, there's yoga. I always pack a yoga mat and candles wherever I go – with these two comforts I know I'm always going to be all right. I'm not an expert yogi but I do enjoy it, although I'm not at the mat every morning because the call times are pretty brutal. If I'm needed to be at work at six, with all the will in the world I don't think I'll get a practice in before then but I do try.

I've been so spoilt on this job because every department is absolutely spot on, from lead director Brian, Jackie the set designer, Lisa in hair and make-up, and Ros with the costumes. They all have an incredible eye for detail, and everybody in every corner of this show really cared from the start, when we knew we had got big boots to fill.

The BBC series is still fondly remembered, and so – quite rightly – there were expectations of ours. This meant that we went above and beyond to get the fabric and the DNA right, and on-set there are artefacts from The World of James Herriot, the museum in Thirsk, giving a feeling of familiarity. When I first emerged in my costume it was like being transported back in time. With the late thirties all around me, I couldn't help but feel I'd stepped into that era.

Yes, I'm spoilt and lucky, and I'm going to put it out there and say I've definitely got the best wardrobe. That's testament to Ros. As we know, Helen was the first woman in her village to wear trousers. I love that! It says a lot about her and that she was forward-thinking, and of course it was practical. I've tried to get on a tractor in a dress during a freezing winter in the middle of the Yorkshire Dales, and it's not pleasant. So of course she wore trousers because it's bloody cold and practical. We worked on getting a wardrobe that was 'practical country chic' (a phrase from Ros). It's lovely, and I would keep all of it if I were allowed.

Through the books we see Helen through the eyes of James, and how he felt about her, but we don't see life from her perspective. Of course, when creating the story in the twenty-first century – and it obviously needs to appeal to a modern-day audience – we want these women to be well-rounded, multifaceted, strong characters. And while there is a limited amount of material in the books, we do know that Helen is based on Alf Wight's wife, Joan.

We were super fortunate to be able to talk to Rosie and Jim, Alf and Joan's children. Through them I gained an insight. They were just little anecdotes but they left me feeling as if I'd peeped behind the curtain. I could understand Helen. Rosie said that if there was ever going to be a cheeky joke in the house, it'd always be Joan who told it. Alf was a little buttoned-up and a bit square, but he would laugh. That sort of anecdote is really helpful because it brings Helen alive for me.

Hundreds of times I've thought about what I'd ask Joan if I could have met her. I'd ask about the relationship between her and Alf. I've heard it from the perspective of Jim and Rosie. They talked to me with such fondness about their mum and dad's relationship, and I almost feel like I've been let into some kind of secret love story. I know they were totally devoted to each other and that they never had a cross word. They had each other's backs through everything, standing up for each other. Their strengths and weaknesses complement each other. I'd like to ask her, 'Was it all as good as it seemed?'

Meanwhile, I hear Helen's voice in my head for much of the year, it's become second nature. She's got that lovely, warm, thick Yorkshire accent, and I can't help but take it on a bit. So I get a little northern throughout that five-month shoot, even when I'm at home talking to my mum.

Helen lost her mum at a young age, is looking after her younger sister, and runs the house ... she does everything. In series two her character develops, and she is having a bit more fun with James as the romance blossoms and Helen's little sister, Jenny, is more involved, and the show is all the richer for it. Imogen is an excellent actress, and she brings real energy. We see the relationship between Helen and Jenny become more like that between friends, as Helen's motherly role fades a little. In one episode we get to put on nice frocks and do a bit of dancing; the set is gorgeous and the costumes are stunning and we had such a fantastic time shooting it. A highlight for me.

Helen and I do have natural similarities. We both love animals (I think they're nicer than people for the most part) and we both love nature, the big outdoors. So it's blissful to spend five months on a shoot, playing Helen. Luckily, I never have to be at the wrong end of an animal because I'm not one of the vets, but if there's an animal to comfort, that's usually my role. Helen grew up on a farm, has been around these animals since she was a toddler and looking after them is second nature to her. And I must mention Clive the bull – a gentle giant. He weighs about two tonnes and is considerably larger than me and, this sounds strange, when I first looked into his huge brown eyes I could see the similarity with my little dog, Rosie. There's an innocence, and I just loved it. (I think Sam may have had his feet trodden on by a cow, but there was no damage done and other than that, no near-misses or scary incidents.)

The welfare of the animals and training of the animals is of such a high standard, and none of us ever feels intimidated because there is always a wealth of wisdom on-set – there's Andy the vet, Jill and Dean Clark the animal trainers, and Jody Gordon the animal welfare expert. A two-tonne bull is pretty intimidating, but there are always well-qualified people telling me what to do.

Working through Covid had its own challenges, with masks on and off, and the social distancing so, at times, it seemed touch and go whether we would get to the end. We heard all the stories of productions that were closing down or having to close for ten days because there was a positive test. I think when we started there was loads of trepidation, no one really knew what was going on. We were coming out of the worst of it, but it was still very much part of the everyday routine – masks and the testing.

We had Phil Pease, our medic and Covid supervisor, who was up to speed on the restrictions and that was a huge operation. However, it didn't stop what we were doing and I don't think it affected the work we were creating. We had to readjust to a new way of working and, once we did that, we soldiered on, just like everyone else around the world, and towards the end of filming the second series restrictions eased and we could take our masks off when we were outside.

The show is about love and community and togetherness, which is so important, particularly in these divisive, cruel times. I feel like I've been so incredibly privileged to be a part of the show and, this may sound a bit cheesy, when I'm not working on All Creatures I really miss it. For me, the show is as it is for the viewer – you just want to be there, be a part of it. It's like a warm blanket.

CHAPTER FIFTEEN

Callum Woodhouse
on Tristan Farnon

For more on haggis, fighting boredom and why he became an actor, Callum Woodhouse explains all.

As an actor, you always want to have a magical story about how you got the part but, to be completely honest, this was just another audition that came into my email folder from my agent.

Actually, I'll be completely honest … As someone pretty new to this business any email about an audition is a magical moment. Acting can be tough. My first role out of drama school was a part in the ITV series, *The Durrells*. When it ended I thought, 'Hollywood here I come,' but I didn't work again for another nine months, so you soon realize how rare great opportunities are.

I'd heard of the BBC series of the late seventies and eighties, but had never seen it. I'd never read the Herriot books (and it was the same when I landed the part in *The Durrells*). On reflection I'm quite glad it was unfamiliar. If I had known how beloved it was, I'd have been a lot more nervous walking into the audition. As it was, I was nervous enough.

Once I had the part, it was a joy to discover the books. A couple of weeks before we started filming, my girlfriend and I went on holiday to Croatia and I packed a few Herriot books, which is what we ended up working from in series one. I thought to myself, 'They were written quite a while ago so they're probably a little bit dated, but I'm sure they'll be a pleasant read.' We were on the sun loungers by the pool and I was getting funny looks every five minutes because I was laughing so loudly. And all the hilarious bits involved Tristan, the guy I was about to play. It just made me so excited to step into the shoes of this character and get to do all the bits that I was finding so funny.

So there you go – just a normal story of an actor not knowing anything about the role, having a really nice audition, wanting to do his absolute best and then, from there, slowly falling in love with the author's work.

There was a particularly special moment. It was after the read-through, shortly before filming, when we met Jim and Rosie. Jim gave each of us a copy of his biography of his father, *The Real James Herriot*, and it's an amazing read. He'd gone through each copy, marking the pages where there are mentions of our respective characters or the real-life people they're based on. Tristan is based on Brian Sinclair, the brother of Donald, who is the inspiration for Siegfried.

In the back of the book, Jim had written some notes about his experiences and what he thought the character was like. This was gold dust. Jim explained how Brian was constantly optimistic, and I was already going down this route with Tristan but to have Jim say it gave me confidence and I was, like, 'Yeah, this is absolutely the direction I need to go in.' He's constantly upbeat, he's constantly up for a laugh, he just wants an easy, chilled-out life where he can have a laugh and get down to the pub. In that sense, I certainly have similarities to Tristan. Life is definitely a glass half full.

And there are other similarities. I love animals, and I think you can tell that from the work I've done, which seems to be exclusively with animals, clearly it's my niche. Then there's the humour. Tristan wants to make everyone laugh so that he is liked. Whether or not he's doing it to hide any insecurities, it all factors in. He's had a tough upbringing, having lost his dad at a young age and in doing so he's also lost his brother, Siegfried, because he tried to become his dad but didn't because he's his brother. So he ended up losing his dad and brother in one fell swoop. He tries to use humour as a barrier, and he's done it for so long that it sort of works for him. That's probably something else that I do. Things may affect me, but more than I let on. I'm just a typical northern man: brush it off and get on with it.

There is a lot of Tristan in me so I didn't have to search far to find his character, but coming back for more series is even more nerve-racking than the first time. That's because the night before filming I have this fear that I won't be able to replicate the role I have played. Oh God, how do I play Tristan again? *How did I do it?* I'll be doing the lines, but thinking that's not how I sounded in the last series and I do not remember how I played him. I used to have the same anxiety shortly before we'd film a new series of *The Durrells*, an awful feeling that I sounded different to the previous series. And then, of course, I get on-set, surrounded by sheep, open my mouth to deliver the first line and it just happens – Tristan comes out of somewhere, God knows where he was hiding.

Why am I an actor? The jokey answer is because I couldn't do anything else. Academically, I am very stupid. I've got emotional intelligence but no normal intelligence. Maybe I'm being hard on myself. My mum always tells the story that she took me to the cinema to see *Babe: Pig in the City*. Apparently at the end of that film I pointed to the screen and said, 'I want to do that.' And then she started enrolling me in acting classes. Maybe she misunderstood me – I might have just been talking about the animal rather than the acting. But film has always been a real passion of mine.

Everyone in my family is into their sports and into their football and I sort of am … but really I'm the only male in my family who isn't obsessed with football. I think my dad realized that quite early on so we'd have film nights every week, or movie marathons where we'd watch, say, all three of the extended *The Lord of the Rings* or the Jason Bourne movies. That became me and my dad's thing, rather than going to watch football. Dad got into films because of me, and then ended up showing me films from years ago that he loves. I've just finished reading *Hellraisers* because Dad introduced me to those classics, and the likes of Oliver Reed's films and Peter O'Toole's.

Then my mum and dad sat me down and said, 'If you want to be an actor you've got to watch *Cold Feet*.' They made me watch series one to five of the original. So from this passion and love I developed a knowledge and understanding of film and television, and a lot of my friends called me the human IMDb because I know all the directors. 'Who's the guy with the black hair in so-and-so,' they'll ask. And I say, 'I can tell you his name for nothing.'

And then, at the age of about thirteen, I watched *Van Helsing* and it became my favourite movie. Hugh Jackman played the title role, and the character of Frankenstein was played by none other than Samuel West. My IMDb had failed me and I'd completely forgotten that Sam was in the film. But I certainly remember watching it and thinking it was the coolest thing I had ever seen. I had all the toys – the merchandise – of the weapons; I had his blades and crossbow. I'd run around my back garden with the weapons, thinking that I was Hugh. Well, the film was added to Netflix and I mentioned this to Sam, along the lines of, '*Van Helsing* – I used to love that film.' And Sam was, like, 'Yes, I played Frankenstein in that.'

'What?'

That night I watched the film once more, and filmed Sam's bit on my phone. Then I sent it to him and on-set the next day I was quoting all of his lines to him, winding him up just as Tristan does with Siegfried.

Within this 'family' of actors, Sam is very much the father figure. Sam talks about things with such passion that he gets me interested. I have never been a birdwatching enthusiast, but Sam is devoted, and when he starts telling me about all these birds I'm like, 'And so what about this one?' I find myself asking him a stream of questions. His love for life is infectious, and he has the ability to totally transport you into his world. No wonder he never stops working!

But I've got a bone to pick about Nick. We have the most delicious breakfasts on-set – in the show you'll have seen us heartily tucking into them – and right at the beginning of shooting series one I said that black pudding was my favourite part of a full English. Nick said, 'You must be a big fan of haggis then.'

I said, 'I've never, ever eaten haggis, never tried it.'

'What, never eaten it? I'm going to cook you haggis the way us Scots do it, with neeps and tatties and whisky sauce.' What a kind offer. Anyway, after all this time, Nick has yet to cook haggis for me. So I've still not tried it, and when I've been offered I've said, 'No thank you, because I'm waiting for the magic moment when Nick makes it …'

Boredom is best avoided on a long shoot. For the second series, I took my PlayStation 5 and Nick took his Xbox. There we were, in our separate rooms and he was on the floor above me, and we played a lot of *Call of Duty*, talking to each other through our online headsets. But from now on my creature comforts will include Ralph, my little dog. I have wanted a dog all my life but Mum and Dad's work schedules wouldn't have been right for it. Making series one I had so much fun hanging around with all these amazing dogs, and then in November 2020 my girlfriend and I got our Ralph. He's a Cavapoochon – a cross between a Cavalier King Charles spaniel, a Poodle and a Bichon Frise. He came up for series two and met Derek who plays Tricki Woo. It's the fastest I've ever seen Derek move as he caught sight of Ralph and he bolted over to him and wanted to say hi. I've got a photo of their meeting – the moment when Ralph meets doggy stardom. Now, it was my girlfriend who picked the name for our dog. I took him on-set and introduced him to Nick, who said, 'Hey! You named him after me …'

We are renaming our fabulous dog!

CHAPTER SIXTEEN

The Legend of Mrs Pumphrey with Patricia Hodge

Patricia Hodge gives an insight into the art of acting, how Derek rates as Tricki Woo, and how she tackles eccentrics.

Here I was, stepping into the long shadows, as it were. I had two predecessors: Dame Diana Rigg, as well as Margaretta Scott, who played the character in the BBC series.

Diana was brilliant and we all adored her. Diana and I never worked directly together but our paths had crossed. Back in 2001 I had presented her with a Lifetime Achievement award at the Carlton Women in Television Awards ceremony. Then – and this was about seven years ago – I was doing a play in the West End and Diana came to a matinee. Afterwards, and very sweetly, she came down to see me in my dressing room. She was extremely generous about my performance, and I was deeply touched by her kindness. I am happy to have that memory of Diana.

Margaretta, however, I did know. On-screen she played Mrs Pumphrey beautifully, although I didn't recognize the Margaretta I knew because that was her interpretation of the character, a performance. That's acting for you.

Before I came to play Mrs Pumphrey I wanted to be clear about what was necessary. It would have been easy for the portrayal to be clouded because there is not just the one previous Mrs Pumphrey, but two. I asked the producers, 'Are you wanting me to reproduce what Diana did?' After all, one doesn't usually take over a part quite this quickly. They said, 'No, you do what you feel is right. Play Mrs Pumphrey the way you think she is.'

I was relieved to hear that because I couldn't have replicated the way in which Diana presented Mrs Pumphrey. Diana is Diana. Each of us has unique characteristics. If you were about to play Hamlet and started to think of all the people who have gone before and their performances … well, you'd fry your brain. So I couldn't allow myself to think about my predecessors. I acknowledged that everybody sees things differently and has a different approach to the character. I just had to read the scripts and let them speak to me. I could and would be my own Mrs Pumphrey.

The approach comes down to what you feel instinctively. I need a clear image in my mind of the character, almost as if I can reach her. There have been times when I have read a script and truly cannot see how to do a character. So it's the writing that's crucial and what it gives you. I am extremely appreciative when the writer provides a little window into the character's soul, and with All Creatures I particularly appreciated that you don't just see one side of the character, you see something that is going on *inside* her. That's much more rewarding, certainly for the actor and I hope it is for the viewer.

In the Herriot books there are the central characters and then those, like Mrs Pumphrey, on the periphery. They dot in and out, and you don't know too much about them. She is a presence, but she's not especially defined. James Herriot doesn't go to great lengths to describe her but, because of Tricki Woo and Mrs Pumphrey's demands on the veterinary practice, she is an important part of the story. She lavishes goodwill and gifts on James, but in the books you don't get an insight into the character. You have to sort of make of her what you will.

When it came to the scripts, however, I was thrilled that we see there is more to her than just the outer layer. It's nice to see that things aren't all perfect in her life because it makes her more of an expansive character.

As the actor, you need to have an innate understanding of where the character is coming from, and you need to have certain elements within you that lend themselves to the character. Ultimately, Mrs Pumphrey is a typical English eccentric. She is one of those people we all know – dotty in a delightful way, and we certainly breed them in this country. They are very much of their own reality, seeing the world in a slightly different way from most people, let's put it that way. Each has an independent view. Mrs Pumphrey sees the world in terms of Tricki Woo, her dog. She is also a woman of considerable means and wherewithal, and is generous hearted. The only thing she needs is somebody to help her with Tricki Woo and make sure he is all right, and that is when she feels extremely comfortable. That's why she loves the vets at the practice – they are the key to her life, her happiness.

For this role, I make Mrs Pumphrey more eccentric than myself. It's a projection, if you like. But I certainly do feel that I understand her. Over the years I have watched all those Ealing comedies of the thirties and forties, and there are characters, eccentrics, that remain with me; little bits of this and bits of that are deep in the psyche. And they are regurgitated, but in a unique way to fit into the character you play. While there is nobody in my life who I would say is exactly like Mrs Pumphrey, I do know I've seen enough characters to enable me to innately know how to play her. That's what I do.

It is my job, but that does not mean there are no nerves. I am often asked about theatre: 'Do you get really nervous before going on stage?' And the answer is yes. And I feel apprehensive shortly before a shoot begins, just as I did with All Creatures. There's never a time when you won't feel apprehensive, but you are at the beginning of creating something and the nerves are built on fear of the unknown. I mean, the number of times that I've gone into an important filming day after just two hours sleep, on frayed edges. This becomes part of what actors live off. We will do what we think is right, but we don't know if it is right until the audience reacts. In the theatre, we know pretty quickly because from the stage we sense the audience's response. In television or film we don't have that immediate response, of course, but we get an idea of whether the chemistry is happening on the set.

Speaking of happiness, All Creatures is such a delightful set to work on. I have worked with Sam on a number of occasions, and while I

Above: Horse master Mark Atkinson. Below: On-set vet Andy Barrett is also Nicholas Ralph's stand-in.

Above: James (Nicholas Ralph) and Helen (Rachel Shenton) early in series one.
Below: Filming series two, with Covid restrictions, James and Helen take to the dance floor.

Above: Siegfried Farnon (Samuel West) with the Rover.
Below: James Herriot (Nicholas Ralph), Helen Alderson (Rachel Shenton) and
Tristan Farnon (Callum Woodhouse).

Above: Director of Photography, Erik Molberg Hansen on set with 'River'. Animals are like the weather, says Erik. 'Both of them can make filmmakers very stressed. But you mustn't get angry.' Below: Director Andy Hay with Director of Photography Annemarie Lean-Vercoe and Anna Madeley.

Above: Callum Woodhouse (Tristan Farnon) with Sheila (aka 'Clancy'), a rescue dog. Above right: Anna Madeley (Mrs Hall) with 'Dash'. Below: The front of Jackie Smith's Skeldale House set, built in what was once a mill.

Above: On-screen father and daughter Richard Alderson (Tony Pitts) and Jenny Alderson (Imogen Clawson) filming series three in the village of Arncliffe.

Left: A break during filming series three. The Alderson sisters Helen (Rachel Shenton) and Jenny (Imogen Clawson). Imogen says, 'My performance has been inspired by Rachel. We have a sisterly bond.' Meanwhile Rachel says of Imogen, 'She is an excellent actress, she brings real energy.'

Right: Siegfried Farnon (Samuel West) and Tristan Farnon (Callum Woodhouse), exterior Skeldale House, Grassington.

Above: On set at Skeldale House for the making of series three, Rachel Shenton and Nicholas Ralph. From their very first audition together, there was that chemistry between them. 'We seemed to find that balance as James and Helen,' recalls Nick, 'and worked well together.'

Above: James Herriot (Nicholas Ralph) and Helen Alderson (Rachel Shenton) in the hamlet of Yockenthwaite, during the filming of series three. Below: A very happy family! The cast and crew of series three, photographed at Arncliffe, June 2022.

didn't know the others they have all been wonderful. I didn't allow myself to feel like the newcomer because, after all, there are guest parts every week, with actors coming to play different characters. And I didn't allow myself to deflect from being inside the stories, being part of the gang. The great thing was that they absolutely made me part of the gang from day one, and without any effort at all. It was a wonderful integration, which I'm extremely grateful for.

I have also reacquainted myself with Yorkshire. About twelve years ago I was in the first stage play of *Calendar Girls*. It is based on the true – and now well-known – story of a group of middle-aged women who posed nude for a very alternative Women's Institute calendar in order to raise money for leukaemia research. The women were all members of the Rylstone and District Women's Institute, and it was set in the same area as the filming of All Creatures. I've remained friends with Angela Knowles, the original Calendar Girl who lost her husband and about whom the story was created. So during filming of the later series I went to see her a number of times.

And what of the real star of the show? I'm talking about Derek who plays Tricki Woo. It has been amazing to work with him. He's extraordinary and the most placid dog I have ever come across in my life, and I've worked with a few. Day one, I was on my way to the trailer and Dean Clark, the animal handler, came up with Tricki Woo – sorry, Derek – and said, 'I gather you'd like to acquaint yourself with Tricki.'

'I'd love to. But I've actually got to get into make-up in a minute.' I held out my hand to Derek, which I always assume you do by way of introduction in order to let the dog get your scent. Derek just looked at me as if to say, 'What are you putting your hand out for?' Then he turned his head away and looked in the other direction.

Derek has not a shred of neurosis and is very, very sweet-natured. Pick him up and tickle his tummy and he purrs like a giant cat. He is also perfect for this show. There are times when, as Tricki Woo, he's meant to be ill and it's as if he takes direction. At one point he was on the vet's table having an injection, and played it like a real pro. You could almost see the anaesthetic slowly affecting him, his eyelids slowly dropping. And that's acting for you.

CHAPTER SEVENTEEN

Playing Jenny Alderson
by Imogen Clawson

The youngest member of the ensemble considers her split
life, in front of the cameras and behind the classroom desk.

I was very young when I fell in love with acting. At Christmas family
get-togethers at my grandparents' home, my cousins and I would
put on performances. Looking back, I can see that I had a knack for
acting and was passionate about performing.

Soon I was at drama school and, within a few years, I had an agent
and next – it happened quickly – I was chosen to play the character
of Jenny Alderson in *All Creatures Great and Small*. My life changed.

My parents are thrilled. My brother Roan – usually my harshest
critic – is seriously impressed and happy for me. (Actually, I think he was

shocked when the first episode came out because he didn't realize how big All Creatures was.) My schoolfriends never miss an episode and think what I do is cool. When my headteacher spots me in the lunch queue he stops to say, 'I love the show.' My school has been really supportive.

I get to go for costume fittings in London with Ros Little, the costume designer. Jenny is a bit of a tomboy, and she likes dungarees. It's odd – I put on the dungarees and the wellies and all of a sudden I'm Jenny and ready for farming. I am devoted to acting and love the show. So at Christmastime nowadays I'm not performing just for the family at home, but being watched by millions, as one of the Aldersons on the All Creatures special.

How did this all come about? Well, this story begins, perhaps, when I was six or seven years old. I always went to drama classes on a Saturday morning and performed in shows. When I was ten I joined Scala Kids, the well-established school of performing arts in Horsforth in Leeds. At Scala you get to act, dance and sing, but Scala is also a casting agency – it finds jobs in entertainment for its performers. The young actors you see on TV shows and soaps, or in movies or on stage – many of them have come through this fantastic performing arts school in Yorkshire. So even if you've never heard of Scala Kids, you'll have been entertained by the talented kids from Scala.

I was just turning eleven when Lynne Walker – the person who set up Scala many years ago – started to represent me. This would be my springboard into professional acting. With an agent, I took a leap forward, not only learning more acting techniques, but also working towards auditions, and learning how to do a self-tape. I had progressed, from those acting classes on a Saturday to having a career as an actor. I was a professional! Or at least I would be, just as soon as I could do well at an audition and land a part. I had no idea how things would turn out and that one day I'd be telling you this story…

Mum got an email from Lynne. 'They're auditioning for a part in …'

'Oh my goodness,' said Mum as she read the email, clearly excited and maybe slightly nervous. 'This is an audition for a part in *All Creatures Great and Small*. Can you believe it?'

Now, of the cast and crew who work on the show, there are those who as children read James Herriot's books, and there are others who remember watching the BBC series in the 70s and 80s, decades before I was born. I had heard of Herriot and All Creatures but didn't know much about it. I was a bit mystified by my parents' excitement. They told me how, as youngsters, they had watched the BBC series. 'It's set in the Dales and it's about a vet and lots of animals.' Then I discovered that my grandparents were also big fans of the BBC series. I was surrounded by Herriot devotees and was like, 'Oh wow, so it's quite famous then?'

When we all talked about the character of Jenny, none of them remembered her. Although that's not surprising because Jenny did not feature in the books. She would be a new character in this adaptation, and maybe a little part of me was eager to make her my own. Lynne had attached a note about the forthcoming audition and it included a character description of Jenny. This explained that she is the little sister of Helen Alderson, and that they live on a farm with their father, Richard. The girls' mother has died. Jenny, said the note, is confident and sassy. There was a short script that I'd need to learn for the audition, with dialogue between Jenny and Mrs Hall, the housekeeper – twelve-year-old Jenny didn't want to go to school and Mrs Hall was trying to get a reluctant Jenny to read a book (later, once I'd landed the part, I was pleased to be told that this same storyline would actually be incorporated in an episode).

This was to be my first-ever audition, and it was taking place in Manchester, about an hour's drive from our home in Harrogate. I was new to the environment, the experience and the feeling. There were others who had come with their parents and, one by one, we auditioned. As Mum and I waited, I was feeling nervous but went through my lines and tried to mentally prepare myself. Next up, I heard a voice, 'Imogen Clawson ...'

In the audition room there were the casting directors, as well as a lady who would read Mrs Hall's lines during my audition. A camera was set up to capture my lines. Suddenly my nerves seemed to vanish. In fact I felt quite relaxed (which, I think, helped my performance to be really natural). I was desperate for the role so I just tried my absolute hardest. After doing the lines in my own accent, the casting director said, 'Can you do it again in a Yorkshire accent?' Thankfully, I'd prepared for this.

After the audition Mum and I were shopping in Manchester when we received a phone call inviting me to a second audition. I was ecstatic and feeling very, very lucky. A few days later we were back in Manchester for that second audition, and afterwards the casting directors told me that I was one of the remaining three candidates for the role. Nail-biting, eh? Then the call – I'd got the part of Jenny Alderson. I was thrilled and celebrated with my family.

Next, preparing for the role. Filming was to start within a couple of weeks. Well, as I have mentioned, Jenny was *new* so I could be the privileged one to tell her story. Yes, I did watch a couple of episodes of the BBC series (I didn't want to watch too many in case I was influenced) and as part of my background research I read about farming in Yorkshire in the 1930s.

Then the scripts arrived. They are written beautifully and, as an actor, the lines give you such freedom to do what you want with the delivery. I read them thoroughly, practised my lines and worked on that Yorkshire accent. I spent a lot of time simply thinking about Jenny's characteristics, her attitude and her outlook. I thought about Jenny's back story; that her mother has died, that she is working hard on a farm and money is tight. I wanted to express the dynamic of being a female, and of being cheeky and confident.

The night before my first day on set I lay in bed, nervous but excited – who wouldn't be? *What will tomorrow be like?* This was my opportunity to begin my career. That first scene was on a farm with Helen, James and Siegfried (otherwise known as Rachel, Nick and Sam). They were walking through a gate and I was in a field and, this might sound strange, but I remember thinking – *hang on, there's a camera*. There we were in the idyllic countryside, and it was all so convincing and then … well, the camera seemed out of place. I'd soon got used to it, of course. From the first day of All Creatures I just knew that everyone would get on and this would be such a special project. (It was comforting that this was also Nick Ralph's first TV role, so the two of us were enthusiastically picking up tips from everyone else.)

Tony Pitts is fantastic as Richard, Jenny's dad, and I have learned loads from him and of how he thinks in advance about the scene. When Tony is acting I can see that he really believes what he's saying, making the performance feel so real. I try my best to do this, too. Meanwhile, my performance has been inspired by Rachel and by simply watching her. We have a sisterly bond and we're similar. We're both quite vocal, powerful women who stand up for themselves.

One of the themes of series three focuses on the power of women. Our voices are being heard. Jenny stands her ground – she won't let anyone belittle her, talk down to her – and she is almost motherly and responsible, looking after her dad and always caring for the animals. From series one to series three, you can see that Jenny has grown up to become an independent young woman with a mind of her own. The development in the episodes of series three sees Helen shift, from being like a mother to Jenny, to being her sister again.

So it is fun to be Jenny, to have some power, especially in the late 1930s (of series three). She is taking on more responsibilities of the farm and becoming a more complex character with a lot on her plate. But she perseveres, and she looks after her sister and her dad, making breakfast for them and doing her chores.

Equally, over the three series of filming, the dynamics have developed between Rachel, Tony and me. On and off set there's a great atmosphere, and we have fun, laugh and get on well. I'll have memories from All Creatures forever. I've had a couple of scenes with Callum and Nick (including a scene in the car in series two). Again, it's a laugh on set with those two, and they always make it fun and check that I'm OK.

With directors like Brian Percival and Andy Hay I've learned so much about TV acting. I remember Andy was directing and I was sort of busying about the kitchen, and he gave me little tips, like think about your mum in this instance, or have a little moment with Rachel, or with Tony. It's advice that helps my performance to be more confident. And when the director is happy, then we're happy and it's a good performance overall.

I'm also fortunate to be filming in the Dales as it is so beautiful, especially when the weather is good. There's a memorable and quite emotional scene in series three, before Helen and James's wedding. Helen and Jenny (Rachel and I) are sitting on a roof – the scene mirrors an earlier scene in series two, when Helen and James sit on the same roof. For my scene with Rachel, it was a cloudy day and the view was incredible.

What about the creatures? At home I don't have any pets but I adore animals so it's exciting to be working with them on the show, particularly Jenny's dog Scruff, played by Bobbi. Bobbi is cute and always happy, so the perfect companion. I'll also never forget Clive, the bull from series one. I took one look at him and thought – oh no, he's the size of a car! I was quite intimidated but then I stroked him and he was fine.

Then, in the finale of series two, Helen's horse Candy gives birth. Candy was played by Aramis, a beautiful black horse. Pauline Fowler, who is in charge of the prosthetics, talks us through how she creates things such as the rear end of a horse, with a lever to flick the tail so it seems lifelike, and the liquid that surrounds the foal when it is delivered. That's what makes it *All Creatures Great and Small*.

In an episode of series three, I clipped a sheep's toenails – a fresh experience for me – and I found myself in a pen with fifteen sheep. Dean Clark, the animal handler, looks out for the animals (he makes sure they get plenty of breaks) and he looks out for me too, making sure we all feel comfortable with a scene. So it might seem pretty scary to be in a sheep pen, but Dean explained how to stroke and feed them. 'Just stay calm and they'll do the same.' He was right. Treat them with respect, and they'll respect you.

How do I manage to fit in my school work? When I first started All Creatures I had just turned twelve years old and at school I had just started Year 8, so there weren't too many pressures. There is a balance between acting and school. I have a tutor on set and my teachers are very supportive and are always helpful when I need to catch up. I must admit there are times when I'm in a maths lesson and dreaming of being on All Creatures. Maths and All Creatures are entirely different worlds! At the time of writing this, it's July 2022 and this month I turn fifteen and finish Year 10. I'm working towards my GCSEs and manage to juggle it all as I love acting so much.

I hugely enjoy sport, and play netball and hockey for my local teams and for North Yorkshire. I like studying, particularly History, and, of course, I'm studying Drama at school. I love watching plays and musicals, and going to London to see a show. My favourite musical is *Hamilton*, which is fantastic, and I'm a big fan of Lin-Manuel Miranda's work. I've been greatly inspired by Emma Watson, who's such a talented actress, and by the story of how she auditioned from her school to get the part in *Harry Potter*. I was also inspired by Sadie Sink's performance in *Stranger Things*. Her acting was incredible and I watch and take tips from her.

So my future seems certain: I will stay committed to acting. As for Jenny's future, I reckon it's on a farm, doing what she loves. Jenny is a strong, independent female and I am honoured to play her. I can't wait to see what's in store for her and the rest of the characters.

CHAPTER EIGHTEEN

Matthew Lewis on Hugh Hulton

Once a pure-blood wizard in *Harry Potter*, and now
James Herriot's rival in romance, Matthew Lewis
is interviewed about life as wealthy landowner,
Hugh Hulton. But first, some background …

When it was announced that there would be a new
adaptation of James Herriot's All Creatures, Mr Lewis
had a word with his son, the actor Matthew Lewis. 'Oh,
you should try to get involved in that,' said father to son. There was
certainly a Herriot connection, at least geographically. Matthew was
born in Leeds, not far from the Dales.

Flash forward, and Matthew – also known as Neville Longbottom in the *Harry Potter* movies – is in New Zealand when he receives the script for All Creatures. Would he have a read and see what he thinks? 'Coincidentally, the make-up artist on the film talked to me about the series because my accent reminded her of it. As well as being an institution in Britain, All Creatures was hugely popular in the Commonwealth. So I was very much aware of its popularity.'

He adds, 'Little did I know just how charming, funny and endearing James Herriot's world really is. Hugh – he's romantically involved with Helen Alderson, making me James's love rival – is the wealthy landowner. There's a degree of deference from the residents of Darrowby. When Hugh walks into the pub, people immediately know who he is, and he's served at the bar straightaway. But he's also warm-hearted, approachable – not the stand-offish lord of the manor who looks down on people with disdain. Although he was born and bred in Yorkshire, he went to a boarding school in London and doesn't have a Yorkshire accent. The key thing is the voice, what Hugh sounds like. It's not an accent I have had to do before, so I worked a bit on that. I watched a lot of *Blackadder*, particularly *Blackadder Goes Forth*.'

The series bible – that document that maps out the episodes, characters and all – describes Hugh Hulton and his part in All Creatures:

When our series begins, Helen is in a relationship with wealthy landowner Hugh Hulton. He's charming, handsome and in many respects a good match – a marriage to him would offer her an escape from the rigours of her current life. The Aldersons lease their farm from Hugh Hulton, which adds another level of complication to their relationship. Helen feels a great sense of responsibility for her family who are under some financial pressure, and she knows that they would do well to keep in good relationship with their landlord, Hugh.

His wealth would bring her family security and he has many admirable qualities that Helen values. On paper he's a good choice but, deep down, Helen may know something is lacking in their relationship, although it's nothing she has confronted … She doesn't recognize yet that there may be other choices she can make and that being a rich man's wife won't ultimately be enough for her.

Cue the love rivalry. 'At times Hugh is quite dismissive of James,' Matthew says. 'He thinks he's quite handsome, but equally thinks, "I'm Hugh Hulton, he's not me!" Hugh is a little intrigued by what's going on between Helen and James, but also believes the idea of anything deeper is absurd. But he and James actually become good friends, and it's a friendly rivalry that they share throughout the series.'

Now Matthew, who lives in Florida, is also an animal lover and says, 'As a child I had two rabbits. Unfortunately, when I left home, I was in no position to own another animal. I was desperate to, but I was constantly working away. Now that I'm married, we have two dogs and live at the back of a nature reserve, so there are turtles, otters, ducks. I reckon there's an alligator back there, as well.'

He relished the chance to immerse himself in Herriot's world. 'With period drama, you have the opportunity of incredible source materials to dip into, and off the back of that there are the costumes and the hair and make-up. By going to a distant place, I find it easy to find the character. It's one of the reasons I got into acting. I was always daydreaming about being in other places, doing other things. When you do a period piece, you're transported to another world, which is a real luxury for an actor.'

Before filming, Matthew was asked whether he had a driving licence and thought, 'Well, that's fine. I've driven in almost every job I've done. And then … "Hang on a minute, this is a thirties car." Suddenly I started to get very nervous.'

On his first day of filming, Nick and Sam were both driving vintage cars, but they'd had the benefit of working with the cars for a couple of weeks. 'So we were able to put the old cars into gear, do handbrake turns … I had this Riley and actually, you know, it was a pleasure to drive. There's one scene where James and Hugh meet on a single-lane bridge. Nick and I were like kids in a candy shop with these old cars, driving up and down, having so much fun.'

And what of the finished product, the show itself? 'It's funny, it's sad, it's charming,' says Matthew. 'Hopefully viewers will find characters that they champion and want to support. There's something for everyone.' And, in particular, Mr Lewis senior.

INTERLUDE

Snack time

The cast and crew of All Creatures are extremely well fed by Yorkshire-based location caterers Daru TV and Film. The business is run by Danny Janes, a well-seasoned former chef, and Russ Kellet, a fine foods wholesaler. Daru's credits include **Poldark, Ackley Bridge, The Secret Garden** *and* **The Confessions of Frannie Langton.** *Here, Danny gives a taste of life as a location caterer … and a few cast members share their thoughts.*

Location catering is not a job for the couch potato. It's a life of long hours for the chefs and the rest of the team as they work hard to sustain cast and crew, all day, week after week. It's not just the vets who need a hearty breakfast to see them through the day in all weathers. Danny says: 'I am a Yorkshireman and very proud of it. One of the successes of any production is how the cast and crew from outside the county are welcomed. I think we have succeeded.'

The schedule

10am: Mid-morning sugar rush and healthy kick. Cast and crew have options that include sticky cinnamon swirl or tray bake. That's the sugar-rush option. Those looking for a healthy kick can have fresh fruit and vegetable crudités with dips. These snacks are delivered to the set, and don't forget the production team at the base.

1pm: Lunch. There are hot and cold choices. 'These are usually served pre-boxed from a Luton van up some country farm track or lay-by, and sometimes there are multiple drops in all weather,' Danny adds. Choices include asparagus and pea risotto, seared tuna loin with salad Niçoise, grilled halloumi with watermelon and caper crumbs. 'We cater for all tastes.' The spinach dahl with roasted cauliflower, for instance, is relished by vegans, vegetarians and those with a gluten intolerance.

4pm: Afternoon tea. This is delivered to the set, with vegan, vegetarian and gluten-free options. And there are pastries galore. Sandwich fillings include egg mayonnaise, honey-roast ham, salt beef with dill pickle and tomato with avocado.

Rachel Shenton: 'The weather in the Dales can be pretty unforgiving so you know the snacks and lunch are a highlight. It can dominate our "family" WhatsApp group. The snacks are great and include chocolate brownies. They were so calorific but

Nick and I always share one because that way it won't feel as bad. Share a brownie to alleviate the guilt!' Fun fact: Over the sixteen-week shoot of series two, the location caterers served 28,613 meals.

Samuel West: eating on-set, off-set and black pudding

- The on-set catering is excellent. I have a regular breakfast, which is two poached eggs, brown toast – no butter – and tinned tomatoes. Paresh, our runner on series one (and our third assistant director on series two) brought my breakfast and it very quickly became the 'setting up thing', which is good because the Dales were cold.
- We shot series one from September to January. We even did our Christmas episode at Christmas, unlike *Mr Selfridge* when we had the Christmas episode in June. It's quite difficult because you forget that you are meant to be cold.
- Every morning I have double espresso in the make-up truck, and several of them. I limit myself to three before midday.
- In the second series we had fantastic pots of individually wrapped portions because of Covid and social distancing. We couldn't eat together, and had to eat on trays. When the sun finally came out, we got to eat outside.
- I rather like black pudding but I did one day forget not to have breakfast when we were shooting a breakfast scene. I had to do about seventeen takes of eating black pudding, at the end of which I had turned into one.

Mrs Hall's Shortbread

Buttery, crumbly shortbread biscuits are an easy-to-make treat. Mrs Hall serves them with a pot of tea … but they are baked by Bethany Heald, food stylist and home economist, who kindly shares her recipe.

Ingredients

250g (8oz) unsalted butter, room temperature, diced

110g (4oz) caster sugar (superfine sugar in the US), plus extra for sprinkling

Pinch of salt

360g (12oz) plain flour (all-purpose flour in the US)

Method

Preheat the oven to 180°C (350°F) /160°C (325°F) Fan (Convection)/gas mark 4.

Grease and line two baking sheets (or baking trays/ sheet pans) with parchment/non-stick baking paper.

In a large mixing bowl, cream the butter, sugar and salt with a wooden spoon. Sift in the flour and gently mix until combined – don't overwork or the mixture will become tough.

Form a ball with the dough.

Evenly roll out the dough on a lightly floured board. Cut the dough into shapes such as fingers or rounds, and use a stamp or fork to decorate, if desired.

Place the shortbread shapes on the prepared baking sheets. Bake for 15-20 minutes – don't let the biscuits turn golden; they should remain pale.

Remove the trays from the oven and leave the biscuits to cool a little before transferring them to a rack to cool completely.

Sprinkle a little caster sugar over the biscuits. Store in an airtight container for up to five days.

Danny's Yorkshire Pudding

Yorkshire pudding is fabulous with any roast meats, the perfect accompaniment for a Thanksgiving feast, and in Yorkshire it is often enjoyed with jam. Go back hundreds of years, and puddings made from a batter of flour, milk and eggs were cooked in various parts of England. But it was the Yorkshire cooks of the sixteenth and seventeenth centuries who gave it a special twist by pouring the batter into a hot pan at the spit, causing the batter to rise and crisp at the edges. Here, Danny shares his recipe for Yorkshire pudding. His version requires equal quantities of the ingredients, using a cup or mug to measure them. 'I've tried many recipes,' says Danny, 'but this is the best one and it's also the easiest.'

Ingredients

1 cup plain flour (all-purpose flour in the US)

1 cup eggs

1 cup equal mixture whole milk and water

Vegetable or sunflower oil for the Yorkshire pudding tray

Method

Preheat the oven to 200°C/gas mark 6.

Fill a cup (or mug) with the flour and transfer it to a large bowl. Make a well in the middle of the flour.

Crack enough eggs to fill the same cup and add them to the bowl. Whisk from the centre outwards, until the mixture has no lumps.

Little by little, whisk in the mixture of milk and water (again, use the same cup to measure).

Next, pour a little oil into the Yorkshire pudding holes in the baking tray and put the tray into the oven for a few minutes.

Remove the tray and pour the batter into the hot oil – careful, the oil might spit.

Bake for 5 minutes and then reduce the oven temperature to 180°C/gas mark 4. Continue to bake for about 20 minutes, or until the puddings are tall, evenly golden-brown and crisp.

PART FOUR

The Creatures, Great and Small

'He's only doing what he thinks is right, it's
all he ever does, even if it's not easy.'

– TRISTAN FARNON ON JAMES HERRIOT

CHAPTER NINETEEN

The Animal Whisperer

(and Her Noisy Home)

Since the early 1980s, Jill Clark has been training animals for the entertainment industry. She is the woman behind many of our favourite on-screen dogs and cats, as well as Shetland ponies, donkeys and cows – and let's not forget Gerry the macaw and Gerald the parakeet. With her son, Dean, she runs 1st Choice Animals, training and handling the animals that have featured in numerous movies, commercials and television series, including *All Creatures Great and Small*. Jill reflects on life, her career, and the pros and the divas we have come to adore.

K nock, knock! Then a third knock. I opened the door to a woman, and she said, 'Can you train a dog to bite a postman?'

A strange question, but let me explain. At the time – and this was back in 1981 – I was running a dog-training club with my then-husband. The Post Office was launching a campaign that highlighted the issue of postmen being bitten by dogs. Being bitten by a dog can be pretty nasty, painful and scarring.

Well, I had a Border collie, a cracking little dog called Jacy, the apple of my eye. So I took on the challenge for the Post Office. I did a lot of obedience work with her and, in time, I had Jacy trained to tug at the laces of a man's shoe. She was ready to play the role. The campaign was a success and, from then on, Jacy's skill at grabbing shoelaces became her party trick. Apart from being the dog that brought me into the film industry, she was the first dog that I could actually call my own.

From there, it has never stopped. For me, the highlight was handling the animals for the opening ceremony for the Olympics in London 2012. It was the best thing ever, but nerve-racking to have the whole world as your audience. At the ceremony I worked with the Queen's corgis and was a few steps away from Her Majesty. It's very difficult to curtsey when you're hanging on to the dogs, who wanted to go with her. That experience won't be repeated – can't be repeated. A complete one-off, once-in-a-lifetime moment. Tough and hard, but fun, and I wouldn't have it any other way.

Things can get very busy. Just recently, our animals have been filming *All Creatures Great and Small*, *Downton Abbey*, *The Little Mermaid*, *Foxfield*, *SAS Road Hero* and *The Darling Buds of May*, as well as a Steven Spielberg production for Apple TV. We haven't known which way to turn.

It's funny how life turns out. As a child, I absolutely adored animals. However, it was my aunt who brought us up, and the problem was that she didn't like animals. We were allowed a sausage dog, Coki, and that was it. He was a beautiful little family dog. Of course, it's a very different home life for me now. I seem to be always falling over the animals, and they're mostly ones that you've seen on All Creatures and other shows and movies.

Almost all of them have human names, from Stuart the cat to Simon the swan, the donkeys George and Daphne and, of course, Derek the Pekingese (Tricki Woo). I have five cats who live indoors with me. There are ten dogs, and they don't come any further than the kitchen, utility room and the yard by the stables. Every morning there I am, in my pyjamas, letting everybody out. At the moment I've got twenty-five geese, and they are in All Creatures and as good as gold. I walk them down to the pond every day, and then every night I just open the gate and call them, and they come scuttling back. Add to that seventeen ducks, along with the chickens, and they all come trundling into the yard. Then there are the deer, goats and lambs, and the horses and cattle. Oh, and a tortoise.

In the mornings you'll find me sweeping the yard and, after breakfast, I start cleaning out, feeding everyone and making sure everyone is fit and healthy. Then I'll take the dogs out for a walk on the beach nearby, followed by a bit of training. It's just constant; there's always something to do.

The Beloved Bassets

While the dogs aren't allowed to come into the lounge as a pack, I do let them in one at a time. I have to have some space away from them at times, although when I do manage to go to bed the Bassets,

Bridget and Belinda, come with me. They are the apple of my eye, can't do anything wrong. You'll have seen them in All Creatures, though they are only in the background at Darrowby – blink and you might miss them.

Bridget and Belinda have interesting roots. Ian Beale in *EastEnders* bought a Basset and called it Chips. Which, of course, was mine. From Chips, I bred Bridget and Belinda from two litters. They are thirteen and eleven, so definitely getting on. They're real characters but very naughty. The way I see it, you have to be a true Basset lover to own a Basset because they are stubborn and defiant. Take them out for a walk and you can be lucky to bring them back again – the nose goes down and the tail goes up and they're off, following a scent. They're off with the fairies. I haven't much longer with them; another year or two, if I am lucky. One has cancer and the other is arthritic, so I want to spend as much time with them as I can because I absolutely adore them.

I do get asked, 'Are there times when you need a break from the animals?' No, is the answer, though I do go on holiday and my daughter will come and look after them. I suppose I'm one of those people who prefer to be with animals rather than humans. My house is alive with animals, but when you start bringing people into it, I think, 'Oh, when are they going to go home?'

My son, Dean, joined the business in 2014. I never thought he would join me, but he did, just like a duck takes to water. He knew exactly how to work with the animals, and it didn't take long before he cracked it. So we are not just mother and son but business partners.

Now, there was a time when I would have liked to be a vet, but I didn't spend much time at school. In fact, I left as soon as I could, and left *home* as soon as I could. So there was no way I was ever going to become a vet because I didn't have the schooling behind me. But I am a devoted fan of James Herriot. I read all of the books, and years ago I was one of the millions who watched the BBC series.

What really appeals to me is the compassion James Herriot showed towards the animals he was working with. He was down-to-earth and, in fact, similar to my own vet. You see, I can sit at the table with him, have a coffee and chat, and we work things through. I'll say, 'My deer's not very well.'

'Well, what's the matter with it?'

'She's … not right.'

And he'll go through it so that he gets an understanding of her temperament. That way we can find out what is the matter. He's a real James Herriot type, taking time with his clients and the animals.

'What Have I Taken on Here?'

You can imagine my reaction when I got this call – it was from the line producer, Tracie Wright, who I already knew. She said, 'Are you interested in doing a new show that I'm working on? *All Creatures Great and Small*.'

'I'm in,' I said. 'I'll do it. I've read all the books. Watched all the shows. Absolutely. Yes. Count me in!' And that was that. Dean would also play a huge part when it came handling the animals.

As the filming got closer and closer I read through the list of things for the animals to do, and thought, 'What have I taken on here?' It went absolutely fine. There was a lot of movement of animals with cloven hooves and you have to be aware, of course, of the rules and regulations.

That aspect of moving animals took a lot of preparation – where is everybody is going to be? But after that it was relatively easy and not too taxing, though it needed plenty of preparation with lots of lying-down scenes and all that giving birth. And I have thoroughly enjoyed it. So have my animals and they're all in the background scenes. Dennis, for instance, is a mischievous little Yorkshire terrier, and a rescue dog. The owners bought a puppy and didn't realize he might pee on the carpet. So he ended up in rescue and then moved in with me. He's full of trouble with a capital 'T'.

Angelica, a Dalmatian, is in the background at the Darrowby Fair. You see her quickly walking past – a proper cameo role. I wanted to call her Angel, but she's far from an angel so she ended up as Angelica. She's completely potty and scatty. Act first and then think about it later – that's Angelica. Everything is at a hundred miles per hour. Dalmatians are very clever, and they think they can get round you by training you first and then get whatever they want. So you need a strong character to stop them from bossing you around.

Dorie, Nemo and Twinkle, our wonderful Shetland ponies, have also appeared in All Creatures. Sadly, I lost Twinkle in 2021. She was so old, in her late twenties, and I had to let her go, but she'd had a good, long life. Dorie came from horse rescue and is still running around on Dean's farm.

And Daphne and George, the donkeys, are still going strong. I bought them years ago. Daphne belonged to someone I knew, and I bought George in Newbury. I'm not sure why, but I did. The thing is, I have a passion for donkeys. My granddaughter has amazing fun trying to ride them.

Patience, Patience

People ask me about the techniques and tricks. It takes time, so this job depends on patience. If we want a cow to lie down, we begin by lifting its leg and then taking it one step further each time. But once you've cracked it, you just have to tap the leg and bingo – she'll lie down. That process usually takes about six weeks, though it certainly depends on the cow. We have one cow that was a little slow, so that took us quite a bit longer. But we managed and she did lie down, she was super.

Also, you need to consider the breed and species. The Jerseys, for instance, are laid-back, placid cows. They are used to coming in and being milked and having that human interaction. So they are much easier to work with than other breeds which have never been handled.

No matter how difficult the animal might seem, you'll always get something out of it. And you can always work around them as long as you know their temperament and basic behaviour. It's a matter of thinking, what is their natural behaviour. Ask me, 'Can you get a cow to tap dance?' and the response is a definite no. They can't do it. Sure, I can get them to move from one leg to the other, but tap dance! The action must be within their capabilities. Working with animals is similar to working with children. It's patience and understanding.

Clickers and Buzzers

I own and live with most of the animals that appear on-screen, and I much prefer it that way because we have a rapport. They know me, I know them and I know what they are capable of and what they are going to do.

Of course, when it comes to animal training things were entirely different in the thirties, when the Herriot books are set. Think of *Tarzan* and the chimps; in those days training was about brute force and founded on ignorance. Now, training is done with positive reinforcement, and establishing the behaviour of each individual species.

These days we clicker-train animals using sound. In the old days, if you'd have suggested clickers, they'd have looked at you as if you had gone round the bend. So it's click, reward (food), click, reward … and the animals get the hang of it. You can actually see the animal start to think, 'Well, if I do that …' I use buzzers when I want them to go from A to B.

We tend to stick to domestic and farm animals, and I don't have a licence to work with wild ones, although training a lion would be very similar to training a cat. It's just bigger and more dangerous!

Oh, and that woman who knocked on my door back in 1981 was called Gladys Hayward, and we have remained firm friends ever since. Gladys bred the cats for Blofeld, the villain in the Bond movies.

CHAPTER TWENTY

The Vet On-set

As the show's veterinary advisor, Andy Barrett casts an expert eye over the scripts of All Creatures, helping to ensure the realism. He's there on-set, and sometimes as Nicholas Ralph's stand-in, performing the real-life procedures such as the delivery of a calf, foal or puppy. This Yorkshire vet has first-hand knowledge of Herriot creator Alf Wight. His first job was at the practice in Thirsk on which Skeldale House is based.

Why be a vet? Well, as a schoolboy I enjoyed studying the sciences, especially biology, and I loved animals, of course. Then there were the farms. I grew up in Worcestershire and had friends who lived on farms. I worked on a local farm and, looking back, think that ultimately motivated my career. So I went to Liverpool University and studied to be a vet.

After graduating in 1988 I applied for a job as a junior vet in Yorkshire. My family is from Skipton, and I was pleased to find myself back in this part of the country. I sent my letter of application to a practice in Thirsk, and was invited for an interview. I knew of James Herriot, and when I was younger the BBC series was essential family viewing. However, I'd no idea that James Herriot was, in fact, the pen name of Alf Wight.

I arrived for the interview at 23 Kirkgate and was greeted by two quite elderly gentlemen, Alf and Donald Sinclair. Halfway through the interview they took me for lunch at the Three Tuns in the square. The meal was occasionally interrupted by people asking, 'Can I take a photograph?' and my white-haired interviewers cheerfully posed for the camera. I'd absolutely no idea why this was happening or why Donald ended up in the kitchen arguing with the chef about the soup. I headed home and decided to chalk it up to experience. Anyway, that evening I was in another pub with a friend, and telling him about the peculiar interview. 'That's James Herriot,' he said. Everything slotted into place.

Much later on I'd discover another side to the story. At the time of my interview, the practice was effectively being run by Jim Wight, Alf's son. Donald was bobbing in and out of the practice and Alf had pretty much retired. But Donald certainly saw himself as the figurehead. Jim had wanted to interview me as well as the other prospective candidates, but he had gone on holiday. 'Leave the interviews until I get back,' he told Donald. Being impetuous, Donald couldn't wait. So he had got on with the interviews, mine being one of them. Jim later told me that when he returned, his dad said, 'We saw a nice lad about the job, but Donald put on a tour de force and probably frightened him off.'

Alf didn't think they'd hear from me again. However, Jim phoned me, checked I was still interested and offered me the position.

That was it: I spent six very happy years working at the practice with Jim, Alf and Donald. For the first six months I also lived at 23 Kirkgate, Alf's inspiration for Skeldale House in the books.

I stepped onto the studio set of All Creatures and instantly it brought back all the memories of working there, living there, being there. The feel of the house was spot on, especially the hallway with the telephone – that really does transport me back to those early days; there's just something about it, and the wallpaper. The little dispensary is incredibly authentic, and that's where we'd gather to chat during the day.

I saw a lot of Alf when I worked there. He'd come into the practice, not only to meet his public but also to take some of the pressure off the practice. There were plenty of visitors to Thirsk, many of them American fans of the books, and they received a warm welcome from Alf. He was a lovely man: extremely friendly, modest and calm, and he always found time to chat. He even gave me a signed copy of a Herriot book that was published while I worked at the practice.

Donald, meanwhile, was an entirely different character. He was – and remains – the most eccentric person I have ever met. Highly unpredictable, quite unreasonable, impulsive and capricious, and he could be pretty bad-tempered. 'Do it this way,' he'd say one day, and the following day he'd tell me to do the opposite. He changed with the wind, and you couldn't be right.

Yet Donald was a charming man and delightful company,

particularly outside work. He'd done so much in his life, was a person with many interests and an ever-flowing stream of anecdotes. He was great to be around, if he'd got time for you. He wasn't pompous, and I admired and respected him. Yes, I liked him immensely. I know this is a little different to how he's portrayed as Siegfried in All Creatures, but I always thought that he was honest about his mistakes. He must have been hugely frustrating to be in business with, but he was one of life's great characters.

I never thought, of course, that one day I'd be asked to be the veterinary consultant on the remake of All Creatures. I'd see Jim from time to time and we exchanged Christmas cards, and then he put my name forward for the series. Jim knew my history with the practice, but I had no experience of TV work. It's been a steep learning curve and extremely different to my day job.

To most vets of my age, and those who have done mixed work with animals, the situations Herriot described will be all too familiar. Alf Wight's genius was to capture the stories and write them down so beautifully. While we still use instruments that Herriot would recognize, such as bloodless castrators, other practices have disappeared. Donald was still using chloroform (anaesthetic) masks for horses when I was in Thirsk in the eighties, but these days there are much safer ways of doing things.

There are some universal human themes reflected in Alf's writing: the emotions concerning loss and illness, and the worried animal owners. Yes, farming is a business and the animals are valuable assets, so there are understandable concerns if anything goes wrong. But it's much more than that. Although the animals are not pets, there is a bond that has been created because the farmer has cared for the animal all its life, and perhaps delivered it at birth. This way of life was superbly depicted by Alf, and is still present today. And just as they did back then, the farmers work hard in a harsh environment.

Some things have changed, though. Yes, we are still often called out to a calving in the middle of the night, but some of the stories show James getting back into the car after a freezing cold lambing or calving, and there's no heating in the car or decent windscreen wipers. Not now. When I get into a car at 2 a.m. at least I can put on the heated seat. We wear Gore-Tex and fleece. The weather can be unforgiving but we have decent kit, comfortable cars and warm houses to go back to.

With the demands of the profession, you've really got to have a desire to do it. Out-of-hours work is pretty anti-social and gets in the way of family life. The profession requires real enthusiasm for the job and the clients, and you need to be able to get on with people. The farmers are straightforward, down-to-earth people. You know where you stand, particularly working in Yorkshire, and I appreciate the directness. You have sensible conversations about what needs doing and what is possible.

When I started at Thirsk nearly all vets in those days were generalists. They were proper GPs, doing a bit of everything: dogs and cats first thing in the morning, then the farm calls, see a horse or two in the afternoon and back to evening surgery.

I moved to Skipton in 1994 to focus mostly on farm work and since then have been doing nothing but that. I enjoyed the dogs and cats to start with, and the horses. But as a vet you realize that you've got to specialize because that's the way the job is now. Specialize in what you enjoy most. And my favourite animals? Strawberry the Shorthorn heifer is right up there. I'm a big fan of the breed, which is making a comeback after they were swept away by the continental breeds that arrived in Britain in the second half of the last century.

I am sent the script, and am fascinated by the process that turns fifty-odd pages into a glossy TV programme. I read it, checking for accuracy, and consider how best we can do the animal scenes. We ask ourselves: Is it right and proper to use animals in this scene or might it be better to use a prosthetic? And can we do it without using an animal? The animal's welfare is foremost in our minds. The animals must not – will not – be put under any stress. That's what a vet's job is about.

Dean from 1st Choice Animals seems to have a pretty remarkable way with the animals. He's able to prepare them well for their roles, get them halter-trained and ensure they are comfortable around people. So the animals are very calm around the crew, and the crew have been respectful. When cows have to lie on the ground, for instance, the crew don't get too close.

When I read the scripts early on, I had some moments thinking, 'How on earth are we going to do this?' To some extent, dogs and horses are quite used to being around people, but farm animals are different. Jill and Dean – as well as Mark Atkinson, the horse master – take time to put the animals at ease. Early on, the producer Richard said to me that as the veterinary advisor I'd have a right of veto; nothing would happen on-set with the animals unless I was OK with it. That's never been a concern because it soon became clear how much the animal handlers care. The last thing they want are welfare issues or to put undue stress on the animals.

The first thing I did was to take Sam, Nick, Callum and Rachel out on a 'boot camp', visiting farms to get them used to being around livestock and horses. These are big, imposing animals and some of them weigh over a tonne. If you're not used to being around them and are nervous then that shows. It must be difficult to give a realistic portrayal if you're concerned about your safety, and I think that the animals pick up on this nervousness … and they don't make things any easier. From time to time the cast must be a bit anxious about dealing with large animals, for example the big Shorthorn stock bull that appeared in series one. It weighed over a tonne.

When I had left Thirsk in the nineties, Donald had given me a couple of little keepsakes: one of his hoof-testers and a hoof knife (he'd inscribed his name into its handle). An early scene involved Siegfried looking at a horse's foot and using both those tools. I thought it would be nice for Sam to have them in his bag, as some of the instruments that Donald had used, and I gave them to him on the morning of filming. He has certainly picked up on some of the key 'Donaldisms'. 'Don't hold back,' I said to him before filming began. 'You won't be able to overplay this part.' Halfway through a cast-and-crew screening of the first episode, I told him he'd nailed it.

It's great to be a part of something that looks so good on television, and to meet the people who do such different jobs and find out how they all work together turning books into a TV series. But it is not always that riveting when they do nine takes of the same scene. Most of us only get one crack at something; on-set it's, 'Let's just do that *one more time.*'

I have acted as a stand-in for Nick when some veterinary procedures are shown on-screen. For reasons of welfare the actors aren't allowed to do these bits, and I'm much happier offering technical advice from the sidelines, and am horribly uncomfortable in front of the cameras. I am a reluctant double.

For episode one of the first series, we had to film an actual calving for a scene with a Jersey cow at Mr Rudd's farm. Most calvings that vets get involved with are quite protracted affairs, but I totally misjudged this one. Having decided that she wasn't going to calve for a while we all went to grab a bite to eat. I was just starting my fish and chips when I got the call to say the calf was on its way. I rushed to the farm and, as I was Nick's stand-in, changed into my thirties vet's costume. I was just in time, as the cow spat the calf out. You can catch a glimpse of my arms, from the elbows down, and I'm glad to leave it at that. It was a nervous moment after nights of waiting and watching for the heifer to calf, and we came within seconds of missing it.

I think there's a theatrical saying about working with animals ...

CHAPTER TWENTY-ONE

Saddle Up

Mark Atkinson is the horse master on All Creatures.
With his son, Ben, he runs Atkinson Action Horses from
their farm in East Yorkshire, with stables of sixty horses.
They train and supply horses for film, stage and shows.
They also teach the stars to ride, and help the jittery
actors overcome a fear of riding. Mark explains his role
on the show and casts an eye over his 'horse-mad life'.

It was the summer of 2019 and we were preparing for that very
first episode of All Creatures. James has applied for the job at
the practice, and Siegfried is taking him on his rounds. They
visit Mr Sharpe's farm, and that's when we meet a farm horse that
has an abscess in his hoof. Siegfried sends James across the mud

to inspect the horse … which kicks James when he tries to take a look at its hoof.

Before filming, Ben (my son) and I went for a meeting to discuss this scene with the director and producer. (These face-to-face preliminary meetings are common, though there weren't many during Covid.) I knew that they wanted a horse to kick James. I'd assumed it would be a hind leg kicking backwards. 'No,' said Brian. 'It would work better for us if it's a front leg kicking forwards.' So that was to be the shot. All we had to do then was establish how to make Brian's wish come true.

Ben watches horses in the field all the time. He'll sit and study them because he wants to see what makes them react in certain ways. A fly lands on a horse's shoulder. What does the horse do? He kicks out in front of him, stamping a front hoof on the ground. This sudden movement gets rid of the fly. So if we could replicate this scenario, we'd have the shot. We have sixty horses, but decided that Aramis would be the best. He's one of our star horses and also appears in the second series, playing Helen's horse, Candy.

So Ben tickled Aramis with the whip so that he thought there was a fly on him, and he kicked out with his front leg. This trick is quite difficult and takes a bit of time, but horses are keen to learn. They buy in quickly; they're thinking, 'He wants me to do something.' As soon as he realized that's what we wanted, the kick would get bigger and bigger. There are no food treats. Once they've done their little trick, we just rub the horse's forehead and praise him. That's it!

Come the day when we had to film that scene, Nick trudged through the mud to reach Aramis. I was about four or five metres away, behind the camera with the whip. A little swish in the air was all that was required – Aramis heard the noise that he associated with praise, and kicked forwards. That simple. Ben always says, 'When you're holding a whip, imagine it as a wand. Hold it as a wand and you'll create magic. Hold it as a whip and you'll create misery.' The whip is merely an extension of your arm.

The best horses will do a trick when they know that that's what you'd like them to do. If they are not very obliging then they're probably not the right horse for the job. Dylan was another brilliant horse, which sadly we don't have any more. He was always super keen to learn a new trick, but would only do it so many times. For a theatre production, he learned to go through a paper hoop, and he was supposed to do that every night. But we knew we'd have to train a second horse to do the same thing because Dylan wouldn't have done it for the whole run; he'd have got bored. You need to know your horse.

All Creatures is a fabulous job. It's fantastic to be working in Yorkshire. It's God's own country, and I am very, very proud of it. The drive takes me an hour and forty-five to get on-set, but I don't mind one bit – what a drive!

I'm also a huge fan of James Herriot and have read all the books. Going back to the mid-seventies, I was a member of a young farmers' club and we'd do competitions in public speaking and prose. I won the junior public-speaking prose competition when I read from James Herriot's *Vet in Harness*. And before she went off to university my daughter, Lucy, did some work experience at the museum, the World of James Herriot. She loved it.

Before that scene on Mr Sharpe's incredibly muddy farm, Nick came to our farm. He was there to meet the horse and to meet us, and have a rehearsal. I think he just wanted to see what he was up against. It's not unusual for actors to visit us. We always push for that if we possibly can. We worked on *Poldark* for five years, and the actors would come and stay locally for three days at a time, but that was a great luxury.

Nick is such a lovely chap to work with, and I admire his enthusiasm and the way he assesses each situation – and he's got so much common sense he should be from Yorkshire! He also has a lot of animal sense. The thing is, I can relax when I'm with Nick, or if I'm watching him, because I know that he'll keep himself safe. With him, there's no bravado.

Can a horse tell if a person is nervous? Yes, I think some definitely can. If your heart is going boom, boom, boom they'll pick up on it. But I would also argue that an actor's horse should be the sort of horse that will help whoever is on it, irrespective of whether they are nervous.

When an actor comes to the farm we'll say, 'Let's have a cup of tea.' We sit with them and have a chat, trying to assess what sort of person they are. Then we choose a horse that we think will match with their personality. If the actor is particularly nervous, we have our go-to horses. We introduce the nervous actor to a small horse so that they can learn to get on and off. It's about building confidence. Next, we'll bring in the bigger horse (the type and size will have been discussed with the director and the art department). We'll have the actor grooming the horse to start with, or leading them around the paddock. Once they're on the horse, I ask them about their career, home life or 'Where are you going on holiday?' It helps to get their mind off riding. Increasingly, we might only have an actor for a couple of hours. You can't teach them to ride in two hours, but you can get them on the horse and hopefully build up their confidence.

Anyone who has driven a car for years has picked up bad habits.

It's the same with riding. When we teach an actor to ride, we try to teach them with just a hint of those bad habits. That way they look like they've been doing it for years. We don't want them to sit up super straight, for instance. It wants to look natural. Then you have it – they sell the scene. And we try to teach them to hold the reins in one hand so that they can use the other hand for acting. Again, it looks natural. I once taught an actor to ride but he was playing the part of a gypsy. Once he'd learned to ride, we then had to change his style of riding so he looked like he had always ridden a horse without a saddle.

There's a sad story in episode three of the first series. Hugh Hulton's thoroughbred, Andante, has colic. James goes to the stables at Hulton Manor to examine the racehorse and then decides that there's no choice – Andante is in terrible pain with a torsion, and has to be put down. (He's made the correct decision, but we have to wait to find out until Siegfried has carried out a post-mortem.)

And when a horse has severe colic he kicks his belly with a hind leg, gets up and down and wants to roll. Playing the part of Andante, there wasn't one horse, but two – Goloso and Ocle. They're almost identical so viewers couldn't tell them apart. Ben taught Goloso to look at his own tummy and to kick at it, but Brian wanted a bit more Hollywood – the horse had to rear up and, standing on his hind legs, spin around in the stables. Enter Ocle. He did a lot of the rearing. (Ocle, by the way, also features in the second series, kicking out at Tristan's shins.)

Morocho (owned by a member of the Action Horses team) was trained by Ben and starred in that same episode, but in a different storyline. It's race day and Siegfried goes to Darrowby Racecourse.

He's eager to land the job of attendant vet there and meets with General Ransom, the pompous, self-important clerk of the course. Suddenly, a horse falls on the track after jumping a steeplechase fence. Ransom says, 'Time to show us what you're made of, Farnon.' Siegfried makes his way towards the injured horse, which is down on the ground, with laboured breathing. While Ransom wants the horse to be winched up in a tractor and removed, Siegfried insists, 'No, the horse is only winded and give him time and he'll get up … Nature takes the time it needs.'

So, Morocho had to lie flat on the ground, beside the actors, as Siegfried ran his hands down the horse's fetlocks, and inspected the horse's eye and gums to check there was no internal bleeding. Morocho lay down for most of the scene, and that's proper acting for you. What a really good day and we are so proud of him. He's done a lot of racehorse scenes since then. Animal welfare, safety and care are foremost in our minds, and it's brilliant to have Andy, the consultant vet, on-set. We look at scenes from the same viewpoint.

Horses have always been my life. When did the passion begin? I can't say because I have ridden since I was tiny. My dad put me on a horse when I was eighteen months old. My son, Ben, won his first rosette when he was twenty-three months old, and Lucy, my daughter, was also riding when she was very little. Then there's my grandson, Charlie. He was born on 17 June 2021, and his first pony was delivered to him a fortnight after his birth.

So you can see how it goes. And then, before me, my great-grandfather bred Hackney horses and was a great horse enthusiast.

Then it skipped a generation, because my grandfather was not that interested, but my dad was passionate; even though he only had horses as a hobby, he was still keen. Naturally, it followed that he taught my older sister, Janet, and me to ride. She had a donkey, I seem to remember, while I had a Shetland pony. Ponies were always there. It was just what we did, and what we got up to do every day. We were surrounded by small farms, but gradually these have been bought up to be included in huge farms. Go back a bit, however, and you'd see big gangs of ponies riding around the Yorkshire countryside.

As a child, I'd compete in imaginary gymkhanas in the garden. My grandmother would play the part of the commentator, introducing me to the crowd, and then I'd ride into the ring, salute and rein my horse back and then put it into canter, fleck it to the right and then jump round the fences. I went to Pony Club, too, and then I took part in gymkhanas – real ones rather than imaginary – and did well. We couldn't buy expensive showjumpers so we produced them ourselves.

I was horse mad and dreamed that one day I would be a showjumper. My hero was Harvey Smith, one of the greatest showjumpers of all time and a true inspiration. And a Yorkshireman. My auntie took me to watch him jump, and I got his autograph. While I watched him in the ring, mesmerized, nerves got to me. Without realizing it, I had chewed the piece of paper in my hands. Only when he'd finished did I see that I'd eaten Harvey Smith's blooming autograph. So I had to get another one.

After leaving school, it was agreed that I would be a farmer, and when Dad retired I took over the farm, though all my spare time was invested in showjumping. One morning I'd finished milking and went indoors for a coffee break. 'Look,' said Jill, my wife, 'you should stop farming. I think we should diversify to horses.' Why? 'Because horses are what make you happy. That's your passion.' She was right.

I opened a riding school and provided riding lessons for the disabled. Then I met some reenactors and got a contract to supply horses for English Heritage and for the Sealed Knot, which performs battles from centuries past. We supplied jousting horses for English Heritage and they paid me to joust. Next, we landed a contract, which we still have, to supply horses to the Royal Armouries for their international jousts; people fly in from all over the world to compete. We supplied horses for a film being made by Birmingham University and gradually we moved into TV work, and films.

Then, one day, 'We need some armoury horses for a low-budget film.' That was *The King's Speech* (with Colin Firth as the future King George VI struggling to overcome his stammer). And then the phone really started to ring. We have supplied horses for television series such as *Jamaica Inn*, *Victoria* and *Peaky Blinders*. We were on *Poldark* for five years and it was as much fun, and as friendly a set, as All Creatures. You can't say more.

INTERLUDE

The 'hero' dogs and cats

There are plenty of dogs and cats in **All Creatures Great and Small.** *Many of them feature in the background of scenes. Others have star status in an episode, and each of these is known by cast and crew as a 'hero' dog or cat. Here's the line-up, with thoughts and observations from production executive Sharon and, of course, animal handler Jill Clark. She knows the animals well as lots of them live with her). For more on Tricki Woo, see page 101, and page 185 for the horses.*

The dogs

Ernie (plays Jess)

Sharon collected Ernie on animal handler Jill Clark's behalf from a breeder in Devon in March 2020, just as the nation was going into lockdown. An eager young dog at the start of series two, he needed a little time to get to know everyone and the set, although it was all very exciting for him. He soon settled into his role as the family dog, and spent much of his time in the studio by block three (producer-speak for episodes five, six and seven). In fact, director Andy Hay treated Ernie like a family dog rather than an actor, putting him in scenes even when not scripted. This resulted in some lovely shots, which enhance the show. Anna Madeley spent a lot of time with Ernie so that they developed a natural relationship and bonded. She sees Jess as a real companion, someone Mrs Hall can talk to when they're alone together in Skeldale House. In one scene, Ernie trashes a game of Scrabble when he jumps on the board. No, it wasn't in the script, but is precisely what a dog would do: scene-stealing improv!

Jill says, 'Ernie is a bit of a diva, full of himself and very clever. He's even learned the word "action". So as soon as you say it he's up and asking, "What do you want me to do? Just give me the right direction and I can manage on my own, thank you."'

Declan (plays Don)

Don, a red setter, is inspired by Alf Wight's own dog. When Alf was a teenager in Glasgow, Don was the family's first dog. Jill says, 'I'd been looking for a red setter to appear in the show, and one day I happened to take some washing to a launderette at a caravan park in Skipton where my son Dean was staying. Well, I walked in and there he was – a red setter. Pointing at the dog, I said to the lady who owned him, "I need one of those." She said, "Oh," looking at me like I was potty. That's how I came to meet Don.'

Frank the dachshund

Frank is the star of episode five in series one, when he's brought to Darrowby Fair with a temperature, and his owner, Mr Happy, hassles James to let him compete in the dog show. Mr Happy leaves the dog tied up by a market stall and … a big, plump marrow falls from the stall and injures him, requiring a dash to the vet.

Handsome Frank was the perfect casting for this role. He's a sausage dog with a strong character, is very obedient and calm. He was also easy for Mrs Hall to scoop up and carry to the surgery at Skeldale House. Frank is a favourite with the cast, especially Nick and Callum who nicknamed him 'Frankums'. He went on holiday to stay with executive producer Melissa and her family. Melissa says, 'He loves to sit on a lap – the minute you begin your descent to sit on a chair, Frank has already launched himself into the air, ready to land. He can jump very high for a dog so small.'

Jill says: 'Frank was given to me by a friend, Roger, before I did the opening ceremony for the Olympics in 2012, and Frank was there with me. At first, I must say, he didn't seem to like anybody, and growled and snapped. I thought, "Oh my God, thanks Roger. What have you done?" But Frank gelled and now he's as good as gold and an amazing, super-nice little dog.'

Sheila (plays Clancy)

Clancy features in episode four, series one, as the gigantic German shepherd brought by her owner Joe Mulligan to the vet to be cured of her vomiting (or 'womiting', as Mr Mulligan says). James Herriot depicts Clancy as an Irish wolfhound, although Siegfried describes her as 'a cross between an Airedale and a donkey', while James considers her as large as an elephant. This huge, scary beast growls ferociously, intimidating James, Tristan and even Siegfried. Mr Mulligan, meanwhile, is unbothered by the growls and barks because he is deaf.

It was going to be difficult to find a wolfhound, traditionally a gentle breed, and train it to do what was required in the time. So the producer Richard and director Andy consulted with Jill and Dean Clark, and the on-set veterinary consultant Andy Barrett. They agreed that a German shepherd could help tell the story where our vets are nervous of examining Mulligan's dog. Sheila could bark to order, and is well trained. If she likes you, she'll bring you rocks – not balls – to throw for her.

Jill says, 'Sheila was a rescue dog from Battersea Dogs' Home. Somebody tied her to the gates of the place with her puppies and just left her. Battersea took her in, and she was looked after for a little while. One of the staff kept one of her puppies. The other one was rehomed, and we then went and picked her up. A lovely dog.'

Ruby (plays Bonnie)

This hugely lovable Labrador was a background competitor in the pet show at Darrowby Fair in series one, and you can spot her in a few street scenes. Come episode four in series two, Ruby is a 'hero' dog in the show, playing in a few wonderful, touching scenes with Siegfried. Jill says, 'During the making of All Creatures, Ruby met and fell in love with Sharon Moran. And Sharon adores her too. So these days Ruby flits between the two of us.'

Kiyo (plays Rex)

Sharon says, 'We needed a very steady, calm and well-trained dog who could play dead for take after take in most of episode six, series two – enter Kiyo.' Kiyo is hit by a car and taken to the surgery at Skeldale House. Sadly, he doesn't make it, leaving a gaping hole in the life of his distraught owner, Mrs Donovan (an older lady from a travelling community, who lives on a barge). Later in the same episode Tristan carries the dog's body, wrapped in a blanket, through a field to the barge, and the dog is laid to rest.

Jazz (plays Susie)

Jazz played the stand-in mother of the puppies in the Christmas special, series one (read the full story, page 38). Jill says, 'Somebody had found her tied up in a cowshed. She'd never been out of it and was terrified of the world. Well, we took her in (she now lives on the farm with Dean), and went on to do the filming. She knows her job, and exactly what she's doing.

The cats

Stuart

Stuart appears in episode one, series two, when we see James Herriot bandaging the cat's paw in the opening sequence. Brian Percival didn't have the easiest day with this particular cat – Stuart didn't like water being dripped on him by Nick when laying the bandage on his leg. As Sharon says, 'It's always the cats – if you've ever tried to take your cat to the vet then you'll understand what a miracle it is for them to sit still for a second.' Melissa adds, 'I've always thought that Stuart looks more like a Bond cat and he perhaps wondered what he was doing in a show about vets …' Of Jill's animals, Stuart probably has the most credits to his name, from *EastEnders* to *8 out of 10 Cats*. 'For *Horrid Henry* we were asked to dye him blue,' says Jill, 'and I've even had requests to dye him pink.'

Jaffa

Jaffa features in episode one, series one, lying on the vet's table – the wrong cat that was anaesthetized. Jill says, 'He was trained to lie down and look asleep. He's an absolutely super ginger cat, and very well buzzer-trained. He's about ten, so getting on a bit, and I've had him since he was about two.' Jaffa's other credits include *A Streetcat Named Bob*.

Lenny and Jet

Lenny and Jet also appeared with Jaffa in the same sequence in episode one, series one – in the confusion over which cat should be anaesthetized. Jill says, 'Sadly, I lost Lenny in June 2021. He came from rescue, and was about seven years old when I got him. Lenny did a lot on *Worst Witch* and then he went on to play Dot Cotton's cat in *EastEnders*. The thing is, I do keep all the animals through old age; they are here until the end. Thankfully, Jet is still with us – *very* still with us.'

'My major animal encounter on-set involved a couple of cats that I was supposed to be able to hold at the same time. One stayed nicely in my arm purring, and did what it was told. The other one didn't want to stay in my arm, and crawled down my back. I tried swapping them over but that didn't help, so eventually we rolled with it – one cat would escape at one point. We let the cats play the scene the way they wanted to.'

ANNA MADELEY

PART FIVE
Making it and Faking it

'Ladies and gentleman, welcome. Today we shall be lifting the curtain on the fascinating world of veterinary science. If I could ask you to step this way, that's it. And please, do refrain from touching the equipment.'

– SIEGFRIED FARNON

CHAPTER TWENTY-TWO

Designing the Characters

'Preparation is the key,' says costume designer Ros Little. 'That means thorough, painstaking research, and establishing a brilliant team that has the right attitude. Good teamwork is fundamental to the smooth running of the department and, ultimately, the show.'

As a junior trainee, Ros, a Herriot fan, worked on one of the final episodes of the BBC adaptation, which was set in the fifties. For the new adaptation, she was thrilled at the prospect of starting in the thirties. 'I love that period and like animals,' she says, 'so I thought it would be great to be involved.' She came on board just five weeks before the shoot began, although that's not unusual in this business. 'Lots of things are decided at a late stage and you have to be adaptable and move quickly to meet the last-minute decisions.'

She began to research the clothes of the thirties, and this would become an ongoing part of the job in series one. 'You need to get it right,' says Ros, who was methodical and disciplined when it came to finding out about the protective clothing worn by vets of Herriot's day. She got in touch with Glasgow University Library, which supplied her with photographs of veterinary students and lecturers from the late thirties. She also approached the Royal College of Veterinary Surgeons, and was given access to the library to look at their clothing catalogues and copies of *The Vet Record*.

Her brother, a vet, suggested she also try the Veterinary History Society, which put her in touch with veterinary surgeons who chatted to her on the phone, and emailed her useful information and photographs. With any production she speaks to as many first-hand sources as possible and, says Ros, 'One eminent vet wrote to me, telling me about his experience when he was a boy in the thirties living on a farm, prior to becoming a vet. It was fascinating and I showed the letter to Sam, which he found very helpful.'

She also pored over her collection of several hundred books 'just to remind myself of what was acceptable in the thirties and what wasn't'. And she went through catalogues of the period, as well as original patterns, as a reference for the clothes made in Herriot's time. 'In remote communities in Yorkshire a lot of people were making their own. They didn't throw away clothes. Mrs Hall and Helen would have made their own, while Mrs Pumphrey would have probably had two dresses, copied from catalogues by a local dressmaker. Catalogue dressmaking was hugely popular.'

As the research progressed, Ros compiled a bible of photographs from the late thirties, both for her own use and for the costume team as reference. Images of farmers and ordinary people in everyday clothing were taken from books illustrating Yorkshire in the thirties.

Some costumes were specially made for the show, while others were bought and hired, especially for the male cast, and often from the London costumier, Cosprop. 'It's expensive to have tailored suits made. You can have the world's biggest budget, but not the time to be able to line up the people to make them.' So Ros used costumes from her own collection, including original dresses from the period. 'They're unfussy clothes, which you can simply slide into,' she explains. 'And I have original patterns, which were also useful. But deadlines were a problem.' She went backwards and forwards to London for costume fittings with the cast, while her team remained on location in Yorkshire.

Ros also worked closely with the producers and the actors had their input. 'Costume design in television is a collaborative process,' she says. 'The costumes had to be a mixture of the truthful and believable, but if the actors wore the same clothes all the time it would be dull for the viewer, so there needs to be a range of styles and colours.' She adds, 'Sometimes what you see is not ideally what you wanted to do, but is the best in the circumstances. There are lots of last-minute changes, and scripts are not available a long time in advance. So when it comes to the necessary requirements for this job, patience is right up there at the top.'

The cricket match in series two was filmed over five days, and brought the usual challenges for Ros. 'There were traditional cricket trousers at Cosprop, but they were woollen and I thought, "This isn't going to work. If the actors fall over on day one it's a problem because costumes have to be washed and dried quickly."' So she had the trousers made in a cotton moleskin that hangs in the same way as woollen trousers, but is quick to wash and dry.

'The shoes were the biggest problem,' she adds. 'The actors needed cricket spikes for safety reasons, but the shoes in costumiers are ancient without a spike in sight. And you can't buy authentic-looking old cricket shoes any more.' She ended up buying pure-white shoes from an online supplier and then covered them in moleskin, the same fabric used for the trousers, which had been dyed to the correct cream for the period. Problem solved, but this process took days; 'Sam wore his old cricket spikes because he plays cricket.'

Ros on the Costumes for the Main Characters

James Herriot (Nicholas Ralph)

Brian and I discussed the colour palette for the character of James, and decided on greys for his opening scenes in Glasgow, developing more colour as time goes on. Then there's that scene, also in episode one, when James – in a suit and shiny leather shoes – is kicked into the mud by Mr Sharpe's horse. For that we made a 'quadrupled suit' – four versions of the same costume in grey wool.

Gradually, we introduced more practical country clothes into his wardrobe, including corduroy trousers, as mentioned by James in the books. He has little money so his clothes were practical and easy to wash; a tweed jacket (which he wears often), one suit, a few jumpers and trousers. It's fun to fit Nick, as he's interested in the clothes, wants it to be authentic and is up for everything.

Siegfried Farnon (Samuel West)

Sam and I have worked together a couple of times in the past, and he is full of ideas about what his character might wear, and when. Before his fitting at Cosprop I'd pre-selected suitable clothes – it's usually a case of the sizes and what's available – and I was looking for particular colours, fabrics and textures. I'd never have anything in the fitting room that I didn't like, and am sure every designer would approach things in the same way. So before Sam tried on the clothes, the choice had already been limited. Then it's a matter of trying on costumes and seeing what works.

Mrs Hall (Anna Madeley)

Anna's confirmation came through quite late in the process. We tried a variety of colours and came up with some that looked very good on Anna, but we all came to the conclusion that she had to look like a member of staff, rather than a woman of the house. She had to look plain and a little bit intimidating when James first meets her.

Anna also has a limited wardrobe, and a work look – after all, she's an employee at Skeldale House. We also talked about her being 'buttoned-up'. Day in, day out, we see her in that cardigan, and it's been mended several times which is perfect – she'd have darned her own clothes. For a special occasion she breaks into colour, and has a beautiful, hand-knitted maroon cardigan. She has a range of overalls, and floral aprons, which Anna loves.

Tristan Farnon (Callum Woodhouse)

Callum loved the jumpers that Peter Davison wore in the BBC series, and he'd found a Pinterest page that was all about the Fair Isle sweaters worn by Peter. Funnily enough, I had some of those knitted in the BBC series. Callum is tall and slim, a bit of a costumier's dream because he's an easy shape to fit and looks good in everything. He's a little bit more extravagant compared to the other characters, and so we wanted to have a bit more fun with his clothes, within the context of him being a veterinary student.

Helen Alderson (Rachel Shenton)

There were quite a lot of suggestions in the script about Helen's clothing. Ultimately, she lives and works on a farm, so often wears practical clothes. James Herriot writes that she was the first woman in the village to wear trousers, with a reference to her green linen trousers. I thought, 'Well, does a vet really know about fabrics? They probably weren't linen.' I found a pair of fabulous, well-worn, green corduroy trousers, and they were a perfect fit for Rachel. She wears these quite a bit when she's working on the farm.

Like other farmers at the time, she'd buy dungarees, possibly from the ironmonger's. So we've tried to blend practical clothes with clothes that are pretty (she has some nice jumpers). She wears an original dress from my own collection. Perhaps her fiancé, Hugh, bought her a few outfits because they're engaged, giving us an opportunity for more variety. For instance, that peacock-blue dress she wears to Mrs Pumphrey's Ball. Stunning!

CHAPTER TWENTY-THREE

Looking the Part

Creating that authentic look of the late 1930s is a craft, and one that requires exhaustive research. The characters' hairstyles and make-up are carefully considered, and meticulously thought through. 'In terms of hair and make-up it's a fantastic project, a challenge, and lots of fun,' says Lisa Parkinson, hair and make-up designer on All Creatures. 'I really wanted to come on board because it's such a wonderful era for hair and make-up.'

L isa is Yorkshire born, raised and based, and for inspiration she turned to the albums of slightly faded photographs of her grandparents. The look for Tristan is based on photographs of her granddad. The hairstyles of some of the female characters have a touch of Lisa's grandmothers when they were young. 'My grandmas had the bowl cuts with the fringe, which was fashionable at that time.'

Lisa is also a Herriot devotee and, as a child, read the books and watched the BBC series. 'That was before mobile phones and laptops, when families watched TV together, and the show was probably watched by everyone who lived in Yorkshire,' says Lisa, whose career began when she was taken on as trainee by Dave Myers, then a make-up artist and now one half of The Hairy Bikers (Lisa was his assistant on the BBC1 drama series Spooks).

She began to research the hairstyles of 1937, and specifically what would have been typical of the working classes at that time. 'Obviously, you're not going to do a hairstyle that's upper class if the character has a working-class background. I didn't want the hair to be too stylized, but reflective of rural Yorkshire rather than the city dweller, yet still look amazing for a modern-day audience. So there's a certain amount of fine-tuning to meet the criteria and appeal to an audience of all ages.' She adds, 'Farming was tough but Yorkshire people are strong. We wanted that to come across in the looks. Farming ladies – and often Yorkshire women – did not have the time or money to have a set at the hairdressers, as many women would have done then. In the evenings they'd rag their hair and sleep in it. They wore a scarf all day, taking it off in public. We wanted to show some looser, rather than contained, hairstyles.'

During her research she discovered the Yorkshire North East Film Archive, a vast library of films – silent and with sound, shot by professionals and amateurs – dating back to the 1890s. One of the earliest films shows Queen Victoria visiting Sheffield in 1897. This archive became an invaluable source of inspiration, enabling her to watch eighty-year-old films about Yorkshire's countryside and inhabitants. 'I wanted natural, normal, and that's what I found in the archive.' She and her team watched hours and hours of Whitsuntide holidays, and footage of the 1937 coronation of King George VI, and the street parties for that day. 'These films and footage take you back to that world. People looked after one another. They were all happy and smiling, even though it was a hard life.'

The hairstyles, she felt, shouldn't be 'too structured' for the series. Hair in 1937 was kept shorter than it is today and rarely dyed. Working women often wore a snood – a headscarf, band or net to protect the hair. And many men wore sideburns, the length of which had to be precise before the actors stepped in front of the cameras. 'The thirties was quite a clean-shaven era,' says Lisa, 'but there were still men who didn't want to shave. The moustache of the thirties had its own style, a full growth but trimmed so it didn't go past the edge of the lips and stopped just under the nose – neat and clipped, like Errol Flynn's, the matinee idol. We made lots of moustaches, including the one for Nigel Havers as General Ransom.'

Lisa and her team made wigs that incorporated a 'finger wave', a style that was in fashion at the time. A hairpiece for Maimie McCoy (who plays Dorothy) took a month to make. Lisa and her colleagues – hair and make-up artists Joanne Tudda and Stephanie Lalley, and hair and make-up supervisor Jackie Sweeney – took turns to thread acrylic hair through lace. 'We made it from scratch because we wanted it to match Maimie's own hair, which was just a bit too short. We made the back piece and then we dressed Maimie's hair over the top of it. It was time-consuming, knotting the acrylic hair one bit at a time.'

Coming up with the look of each character was a collaboration between Lisa, Richard Burrell and Brian Percival, and the actors. Lisa also had to consider the weather and its effect on hair. 'We're known for rain in the Dales so we have to do something that will keep up well in the weather, otherwise it can be a different look by the end of the shoot.'

Lisa was concerned for the child actors because, ideally, they'd need quite a dramatic haircut. 'Children have very long hair nowadays so that could have been a challenge. When we approached the supporting-artist agencies I'd asked our second assistant director John Turner to tell them that the hairstyles are quite short for All Creatures – would the children be happy to have their hair cut? Most happily agreed, and there's only been the odd occasion when we've had to use a wig or cheat by pin curling or plaiting.

'We had one great group of girls. Rather than being upset about having their hair cut, they decided to donate their shorn locks to charity. One of them let us cut off seventeen inches, and then donated the hair to be made into a wig for children with cancer. So a lot of the other children did the same, and through the cast and crew we raised nearly £450 for the Princess Trust.'

Lisa on Styling the Characters

James Herriot (Nicholas Ralph)

He has such a strong presence on camera, and we needed to enhance that. We came up with certain looks that we thought might suit him, and tried them for the camera tests – that's where we try out our cameras and lenses, and the actors wear their costumes and are in hair and make-up so we can review their look. Nothing seemed right for Nick. He always looked too modern, and his hair has a lot of colour flecks that would've picked out on camera. Finally, I gave him the slicked-back look and it was quite a transformation. The style suits the period and Nick looks like a screen idol. Perfect! We all agreed that was the look. We used wax, but then Nick rang me to say he couldn't get it out of his hair. It's not water-soluble so it doesn't shift easily. I told him to use neat washing-up liquid, and eventually the wax broke down and washed out. But we realized that clearly this wasn't going to work over a long shoot, and changed products.

Siegfried Farnon (Samuel West)

Coming up with Siegfried's look was a collaboration between Richard, Brian Percival, Sam and myself. We didn't want anything too harsh for him; while beards weren't fashionable at the time we felt it gave him a softer look. It's a full beard but a little unruly. All the main cast members were given this 'softer' look.

Mrs Hall (Anna Madeley)

Anna is incredibly beautiful so very easy to make up. We had to colour her hair because she had blonde highlights, and went for a style that was professional but still soft and made her approachable. We did quite a few looks together, and quite a few wigs – rolls, curls, waves. We ended up with Anna using her own hair and colouring it. Her first day of filming was also her fitting day – it was a bit of a rush!

Tristan Farnon (Callum Woodhouse)

Tristan was an easy look. Callum has a lovely, natural curl in his hair and is very handsome so obviously he can pull anything off. As he has that natural curl I wanted him to have a softer wave. I looked at my granddad in those photos and that's what I decided on for Callum – a curl at the top, short on the sides. Lots of men had that look in those days. Times were hard but men, farmers included, usually wore suits, were clean-shaven and their hair was quite neat, much more than I had thought before researching.

Hugh Hulton (Matthew Lewis)

With an aristocratic background, Hugh had to look different to the working-class characters. He needed to be sleek. Before series one, Matthew had been in the States and his hair had grown very long, but he was up for anything that was necessary for this thirties role. So we cut his hair – a lot – and on-screen he is clean-shaven; very different to the Matthew who arrived with full beard and hair down to his shoulders.

Helen Alderson (Rachel Shenton)

We wanted to make sure that she was seen as a modern woman, but also a strong Yorkshire person. She looks after the house, the family and the farm, and she's a very attractive lady. Rural women weren't always in their dungarees and looking a bit shabby. They did dress up and they had their Sunday best. We wanted both looks for Helen, and so got to play with different styles, including the snood for her hair, and also with her hair free and long when she's out socializing.

CHAPTER TWENTY-FOUR

From Kings to Cows, Making the Role Models

There are times when some creatures, no matter how great or small, are required for a scene but for one reason or another cannot be used in filming. Only a model will do. That's when prosthetics are called for, and producers often turn to specialist company Animated Extras. With a few decades of experience, and silicone always at the ready, Pauline Fowler is the model-maker extraordinaire whose lifelike animal replicas play a key part.

'Can you make me a cow's neck, please?' The producer Richard was calling to offer me a job on All Creatures. I'd met him in 2008, when we worked together on *Silent Witness*, the BBC crime series about a forensic pathologist, played by Emilia Fox. My response was a resounding yes. That neck was for episode six of series one, and there were plenty more prosthetics needed afterwards.

You can see our work throughout the three series of the show, though I sort of hope you won't be able to tell the fakes from the genuine, live ones. At Animated Extras we strive for lifelike models, as real as can be. There are the prosthetic puppies in the Christmas episode of the first series. Thea Mulvey, a freelance sculptor, created the 'hero' puppy, which we cast in soft silicone. Then I had to hand-lay the hair with tweezers as the puppy was so tiny. I also made another four or five little puppies, fabricated by hand from scratch, using matching fake hair. What great fun! Then there have been the bums, created for the birth scenes: we had a sheep giving birth, plus a horse's bum, complete with birth canal and a flicking tail.

One scene meant recreating the rear end of a Jersey cow, Milo, because she was calving. Now, for that one we were lucky. A long time ago we had worked on *Band of Brothers*, the hit mini-series about a company of US paratroopers, set in the Second World War. We had sculpted a Jersey cow, which was used as a dead beast on the battlefield. Having been in this business for three decades, we have an extensive archive and have accumulated an awful lot of fibreglass moulds. They are durable and long-lasting. So if ever someone comes along with a request, we see if we can make it work with one of our existing moulds.

For Milo, we got out that cow mould. There was a minor challenge because we had to re-jig the bosoms: those from our existing mould were big, while Milo was slightly more petite in the bra section! I readjusted the size, luckily having made the udders with a soft polyfoam. Then I made a soft birthing canal for the actor to insert his hand into, making the experience a bit more pleasant for him when playing a vet.

Another scene required a stomach operation on a Belted Galloway, the Scottish breed of cow that's black with a distinctive white 'belt' running from around its back and under its belly. I used the complete belly and back leg section from our existing cow mould – it's a multiple-breed mould! We began with a fibreglass model of the cow's body, down to the legs. A large silicone insert of softer skin was then put into the side of the body to be operated on. The section of silicone was hand-punched with hair, and clipped back so that it really did look like shaved skin (one thing about silicone is that hair doesn't stick to it). Finally, I had to make a large silicone stomach, which had to be pushed through the incision. Cows' stomachs are huge!

Sam, as Siegfried, had to make an incision with a knife. During the scene I was on the other side, crouching out of view. I had to push the silicone to create a lifelike representation of the stomach moving as the procedure was performed. I sat on one side pushing it out, and Sam pushed from the other side. After the director said, 'Cut,' I said, 'Sam …' And then I stuck a lollipop through the incision. Sam burst out laughing. I'd thought of taking a hand puppet, but decided a lollipop would be more fun.

I was also able to share a little story, a memory, with Sam. Way back in my early days, in the mid-eighties, there was a television series called *Mapp and Lucia*, based on the novels by E. F. Benson. Sam's mother, Prunella Scales, played Elizabeth Mapp. At the time I was sculpting waxworks, and was asked to create a life-size model of Prunella (as well as her co-star, Geraldine McEwan). It was for a display at the Ideal Home Exhibition. Sam was chuffed to hear that. So from waxworks to lollipops, and a big circle for me and their family.

Working on the series meant travelling to Yorkshire, which I love and know well as my parents lived in the market town of Keighley. Mind you, the first three visits to the set were not so great. It was absolutely freezing. Milo the cow was in a nice, warm barn. I was outside, shivering in temperatures of minus nine, painting the replica of a Jersey cow. Andy, the vet, felt so sorry for me that he gave me one of his hats. And he helped out sewing stitches onto the prosthetics – he's a vet, after all. He certainly knows how to sew.

'Can You Sculpt a Hippo, Please?'

Here I am, in the middle of my workshop at Shepperton Studios. On the walls around me there are white plaster faces of actors and actresses that we've taken casts of over the years – a who's who of the greats, from Helena Bonham Carter to Tom Hanks. Directly in front of me is a medical gurney bought on eBay with a disarticulated skeleton stretched on it. If I turn my head, there's a full-size polystyrene African elephant head suspended on our mezzanine.

The workshop smells nice and fresh because we've just done a deep clean. Normally, however, there's a heady mix of all the materials we use, including clay, plaster and acrylic. It's pretty quiet at the moment, but come most days and there's the buzz of drills, extractors, milling machines and swearing. I am currently working on some mummified bodies. I like to create dead bodies, and skeletons are also enjoyable: they don't talk back. One mummified body is going out this afternoon, to be filmed. Then I'll start the next one. Yes, this job keeps you on your toes.

I am thinking back to where this all began. How did I end up in this business? When I was a child, I wanted to work at London Zoo, but for most of my childhood and teens I didn't have a clue what to do. But when it came to art, I was good with clay. I worked my way through a BA degree and got a first. Then I studied at the Royal College of Art, where I did a Master's degree in ceramics. But I was still unsure where this was heading. I remember telling my tutor, 'Whatever happens, I don't want to be just an artist in a garret. I want to use my skill to make a living.'

I held an exhibition of my works, which was great. Then I held a second, which was rubbish because, I think, I simply didn't know enough. But then I met a mould-maker who said, 'Would you like to be my sculptor?' It seemed exciting. I didn't sleep for a couple of years because of the demands of the job, but in the process I discovered that I had a facility for the job. I could do it!

A few years later, when I had moved to another company making waxworks, my career veered towards film and TV. My boss got a phone call: 'Can you sculpt a hippo, please?'

'No,' she said. 'But I know someone who can.' She meant me.

This was the assignment that introduced me to the people who, ultimately, would become my business partners: Nik Williams and Dan Parker (who has since gone solo, lives in France and is doing extremely well). So Nik and I have been partners for about thirty years. He is much better than me on film work. Meanwhile, I love TV work. Why? It's quicker – there's no time to get bored.

Everyone at our company has a different skill set. I am quite good at the guts, the blood and cutting things up. Then there are the skin painters. This is an amazing talent: painting translucent layers of liquid silicone that's mixed with oil to slowly create lifelike skin. Often people think that skin is translucent, but it's not. It requires working with an opaque base and building up the depth through those micro-layers of the liquid silicone; a gradual process but worthwhile because the result is incredible.

I particularly like to work with silicone. As a sculpting material, it's exciting – so organic and lifelike, and with it you can achieve the softness and the texture of skin. Plus, actors react to it because it feels real.

Crowning Glories

For the Netflix series *The Crown*, we made the body of King George. He'd undergone an operation on his lungs, and was later embalmed. So we made a replica of his body. It was a fascinating job, and not least because we worked closely with the surgeons at Guy's Hospital in London. Stephen Daldry, the director, asked the surgeons to perform the operation for the cameras. So there were real-life surgeons operating on a prosthetic king. Apart from the lungs, I made arteries that could be tied, as well as a pumping heart. The surgeons were fully engrossed in the operation, so much so that they kept moving in front of the cameras, obscuring the view. I'd hear, 'Move away from the cameras …' And then a minute or so later, 'Please – move *away* from the cameras.'

For the movie *Ammonite*, we were asked to create a life-size replica of a young whale, about three metres long, which was beached for the scene. The whale was designed in 3D and then printed out for us in multiple sections in polystyrene. Then it was brought to our workshop. Clay was applied to create the skin and add the detail. Finally we moulded it, cast it and put the sections together. As you can see, that's a lengthy process and took about eight weeks. There are no short timeframes.

On the whole, every job is different. Each has its own requirements, which make you wonder, 'How do I work this one out?' But it's fun finding solutions. So we are lucky; and while we might not be millionaires we've had a blooming good run, along with meeting incredible people. As we all know, there's a mass of talent across the board in the film business – not only the stars, directors and producers, but the people who provide 'the stuff', from the props to the tech designers. There is a huge skill base in this business and – amid the life-size polystyrene elephant, the plaster faces of Tom and Helena, and the skeleton on his gurney – it's great to be a part of it.

The Buttercup Mission

'Can you make a boil for a pig's ear, please?' This was a request for an episode in series two, and another first for me. The storyline focused on Buttercup the pig – being played by Flamingo – having a nasty boil on her ear.

I was unable to go and meet 'the actor' because she was in Yorkshire and I had lots to do at the workshop – the other animals were keeping me quite busy. So George and Sarah from the All Creatures' art department photographed her and measured her ear. I knew the colours and dimensions. Next, I had to find a lookalike pig that lived near me (the animal handler Dean helped me out by locating a local pig). Yet it was still guesswork.

I mocked up an ear shape in clay and moulded it. Then I sent a few pictures of clay sketches of large boils to the production designer Jackie. She chose the size she liked and I moulded that one, and then cast it in foam latex. Finally I painted it and hand-laid some very coarse black hair on it as an edge blender. Pigs' ears are incredibly hairy, with more of a yard-broom texture than hair.

I had to make the back of the appliance pre-sticky, and then press it down on Buttercup's head … and pray! Because of the stiff hair it floated about a centimetre above her skin like a red hovercraft. But luckily it did stick well.

That's the thing about working on *All Creatures Great and Small* – always fun and never a dull moment.

PART SIX

Roll the Credits

'It's not the mountain we conquer, but ourselves.'

– MRS HALL

CHAPTER TWENTY-FIVE

Cue the Music

Alexandra Harwood had been struggling with the theme tune for All Creatures. Then, one morning, she put a lead on Brinkley, her golden retriever, and – 'Off we go!' – they went for a walk. When they next came through the door the composer would have created the theme tune for the show. The entire melody had come to her as they strolled in Old Deer Park in Richmond, south-west London.

'We were in the middle of our walk,' Alexandra recalls, 'when in popped an idea. Ideas come to me when I am walking with Brinks, and this was a melody. I worked that melody over and over in my head, thinking, "Can I remember it, do I like it and is it still there?" If it disappears quickly then I realize it's probably not good enough because any really good melody should be very memorable; it's not just whether one likes it or not. I thought, "This one might be it."' In the middle of their walk, she stopped, took out her iPhone and sang and hummed it into the voice recorder. 'Dah-dah-dah … '

'Sometimes I'll get home, play it and go, "Oh no, that's terrible." But on this particular occasion I got home and played it on the piano. Then, once I'd perfected it, I made a demo in my computer program and sent it to the producers. Thankfully, everyone loved it.' This minute-long melody became the title music, shortened to the animated title sequence of thirty-three seconds.

Alex has a studio at her home, with a grand piano, a computer and MIDI keyboard which uses sample libraries (with sample instruments) to create music. She can 'pull' sounds to create any score. Every articulation of each instrument is available. 'I can write with bowed strings, plucked sounds, wind instruments, electric sounds or even the whole orchestra at my fingertips. Everything is there, and I can play using every instrument I want.

'Having composed and played the piano since I was four years old, I think my fingers have a muscle memory. My fingers and my brain are linked somehow without me being conscious. When these melodies come into my head I might be sitting at my desk, with the film on-screen in front of me, and I'll just be improvising to see what comes out. Inevitably something comes out that's generally what I am after. It may not be perfect, but it's enough of an idea that I can work with.'

Does inspiration come from sitting at the piano? Or does she sit, ruminate and muse over a glass of wine? 'Kind of all the above. Though it's more likely coffee than wine! With All Creatures, generally what happens is I might read a script and I might hear the actors' read-throughs, which I did more with the later series than with series one. But then I wait to be sent the edited picture of an episode, and then either sit at the piano or fiddle at the computer on the MIDI keyboard.'

Frequently, however, 'the best things happen when I am not really trying to think. I remember Dad, who was a writer, getting really cross with us if we disturbed him. And I now understand why. If I'm walking around with a cup of tea the kids might think, "Mum's not working right now." But often that's when everything is quietly working away in the background of my mind.

'In bed at night, I'll wake up with an idea. Last night I went to bed a little bit low because I was really stuck on a cue. I woke up this morning in a good mood because I'd solved it in my sleep. I went to the computer and played it, and thought, "Oh yeah, that worked, thank God."'

For All Creatures she created themes and motifs for each character, and some of them have an individual instrument that she felt lent itself to their part. 'James's theme is synonymous with the main theme of the Dales, as I felt the Dales and James are the central characters of the story and somehow related. Siegfried's theme is played on a bassoon, which seems to embrace his humour and seriousness at the same time. Helen's theme is either played by the piano and the oboe, or just the oboe, and has a motif accompaniment in the strings and harp. A variation on this theme is the love theme between Helen and James. Tristan's theme is made of a cheeky clarinet melody and pizzicato strings for when he's up to mischief. Mrs Hall and her home share the same theme as I felt she is the steady centre and comfort. And Jenny's theme is played by the flute, which I felt has the lighter innocence of a young girl.'

As for Mrs Pumphrey, Alex says, 'She has an orchestral theme, which is a regal variation of the Dales theme, and a second, melodramatic theme for her more dramatic and emotional moments, especially when she's concerned about Tricki Woo. And last but not least is Tricki Woo. Even though he's a little dog, I felt he has a very big character that could take a low contra bassoon and a very distinct comic theme.'

Alex was born in London in the sixties and 'grew up with the arts'. Her father was Sir Ronald Harwood, the playwright, author and screenwriter. His film credits included *The Pianist*, for which he won an Oscar. Alex's mother was Natasha Riehle, a ballerina who became a stage manager. 'Music was on endlessly in our house,' she recalls.

At the tender age of four she composed her first piece of music. At school – Bedales, in Surrey – she was writing musicals from the age of seven ('The school put them on, with orchestra, the whole thing'). She studied classical composition at the Royal College of Music, followed by a postgraduate year at the University of California, Los Angeles, and then took a Master's at the Juilliard School, New York's acclaimed music conservatory.

While in New York she met and then married an opera singer. Then, at the age of twenty-eight, she decided to end her career. 'I stopped composing. I wanted kids.' When it was discovered that Harrison, her eldest son, was suffering from a kidney disease, the family moved to London so that he could be treated at Great Ormond Street Hospital. 'The doctors there basically saved his life,' says Alex, but the treatment lasted twelve years. Two more children followed, Phoebe and Joshua. But because her husband was away typically ten months of every year, inevitably the marriage deteriorated. 'We had grown apart in a typical if-you-are-not-together way.'

When the children got to their early teens, needing her a little less, Alex began to feel something was missing and incomplete without composing and music in her life. 'I felt my past musical life that I'd turned my back on was tapping me on the shoulder saying, "Don't you forget your roots."'

She taught music theory to pupils at St Paul's School in London, and then began composing for short animation films for animation students. This led to her taking a second Master's degree at the National Film and Television School in composition for film. Alex was making a name for herself (her credits include Disney's *Growing Up Wild*, *The Escape* and *Thatcher: A Very British Revolution*) when the director Mike Newell asked her to compose the score for his film *The Guernsey Literary and Potato Peel Pie Society*.

In the summer of 2019 she received a letter. 'It said that our vet, Simon, was leaving his practice. He's brilliant and also the nicest person in the world. In the months leading up to his departure, I'd had to visit the practice more than usual as my old black Labrador Milly was getting very ill, and we'd just got Brinkley as a puppy. I always felt Simon and I chatted so easily. I felt like I was losing a close friend.'

Some months later, in November, she was close to finishing her musical composition for Northern Ballet's *Geisha* when she received an email from Simon asking if she'd like to meet for a drink. They met and then, a week later – and after another examination of Brinkley's eye – they went to a café. Over tea and cakes, their conversation went like this ...

Alex: 'Tell me about your early life as a vet. How did it all begin?'

Simon: 'Well, I went to Liverpool University and after that it was a bit like the old TV series *All Creatures Great and Small*. Do you remember that?'

Alex: 'Oh, I remember that.' And then they chatted about the BBC series, which she hadn't thought about since watching it as a child in the seventies. Their friendship was developing, albeit quite timidly, towards romance. Two days later Alex received a phone call from her agent: 'They're making an adaptation of All Creatures. Would you like to pitch for it?' As Alex says now, 'I mean, what were the chances?' She sent a message to Simon: 'You won't believe it …'

Alex was appointed as the show's composer. As for the romance: 'I think we went on six dates and then finally got together. My life has been vets and animals ever since. I was scoring for vets and animals, and I was dating a vet and talking about animals the whole time.' With a kind-hearted vet and romance, the story certainly has the ingredients that would have enchanted Alf Wight, whose great passion was music.

There was another delightful twist when Alex told Simon that a man called Andy Barrett was the veterinary consultant on All Creatures. 'Andy?' said Simon. 'He was in my year at Liverpool.'

'So I reconnected them,' says Alex. 'Above his loo, Simon has a photo of his university graduation. There they are in the picture, Simon and Andy, together. Life is ridiculous, isn't it?' Brinkley, therefore, takes the credit not only for the union of Alex and Simon, but also the reunion of Simon and Andy.

The Covid pandemic brought its own struggles for Alex. *Geisha* was on stage for only one night before the curtain came down. 'We were supposed to have a twenty-eight-day run, including Sadler's Wells. We had the opening night and then all the theatres shut down,' she says. 'It was devastating.' With All Creatures, 'we were really lucky as the filming for the first series had just finished before lockdown. I don't think it would have gone ahead if they hadn't finished.'

Covid also reunited Alex with her daughter, Phoebe, who returned from New York, where she was living and studying. Phoebe's journey was serendipitous. While back at the family home she began to help Alex. The American episodes are five minutes longer than the UK episodes, and the UK versions have commercial breaks whereas the American episodes do not. 'Phoebe started calculating the differences between the American and UK cuts. She was telling me where things would need changing. She'd say, "Well, there are four frames out there, five frames out here. There are twenty extra seconds here …"' Phoebe mastered the audio software Pro Tools and Alex set her up on another, second desk in the studio. 'Effectively she was doing the job of a music editor, and realized it could be her career.'

So, Phoebe was music editor on series two and three of All Creatures. Alex adds, 'She's a really good emotional support for me. It used to be my mum who I would play things to. She was my "mirror". Now it is Phoebe who's become my mirror, and I play to her what I've composed. But deep down I know that if I need someone to listen, so I can read their reaction, it's because I know it's not right. When I know it's really working, I need to do that less. But I do bounce things off Phoebe and rely on and trust her.'

She adds, 'Dad once gave me the best piece of advice when starting on a new idea or tackling writer's block. He said, "Not only give yourself permission, but make yourself put it down *badly*. Then you have something to go back to and work with." I think we can cripple ourselves with perfectionism – trying to make ourselves put down those first ideas perfectly. Dad's advice is liberating and it's actually fun to make yourself do it badly. And that often leads to inspiration.'

These days, 'one of my favourite moments to score is when people are about to kiss on-screen'. Why? 'There's that feeling of getting inside their heads – what does it feel like just before you kiss somebody you really like? If I can make myself get a bit goosebumpy as I write the music for the scene, then I know I've done my job. If it does nothing for me it can't possibly do anything for anybody else.'

Alex says, 'Music can do so many different things for film. It can be another character in the story or tell the audience something that is not on-screen. It can help us be inside the head of a character, and help us understand what they are feeling, like James's feelings for Helen. You can imply so much through music and help the audience feel more, without them being aware of it. Music can help us feel something when we don't necessarily see it.'

Key Locations

Skeldale House (interior)
A studio set near Harrogate

Skeldale House (rear exterior)
Arncliffe Village

Skeldale House (front exterior) and Darrowby, village
Grassington

The Drover's Arms (exterior)
Devonshire Arms, Grassington

The Drover's Arms (interior)
The Green Dragon Inn, Hardraw

Herriot household, Glasgow
Bradford Industrial Museum, Bradford

Waterfall (James swimming)
Janet's Foss, Malham Lings

Mrs Pumphrey's mansion
Broughton Hall, Skipton

Railway stations
Keighley (Glasgow Station) and Oakworth (Darrowby Station)

G. F. Endleby, grocer
The Stripey Badger Bookshop, Grassington

Glasgow Streets
Cater Street, Bradford

Benson's Farm, Yard and fields
Kettlewell

Darrowby Village Hall (Daffodil Ball)
St Wilfrid's Church Hall, Harrogate

Sebright Saunders Estate / Sennor Fell
Sawley Hall Estate, Ripon

The Renniston
Ripley Castle, Ripley

Ritz Cinema (exterior)
Westgate, Thirsk

Canal towpath (exterior) and Mrs Donovan's canal boat
Newton Grange, Skipton

Pumphrey cricket pitch, rear lawn and marquee
Studley Royal Cricket Club (second pitch) Fountains Abbey Deer Park, Harrogate

Ministry of Agriculture
Crescent Gardens, Harrogate

Hulton Hall and Stables
Norton Conyers, Ripon

Alderson's Farm (exterior)
Yockenthwaite Farm, Hubbersholme

Churches (James & Helen's Wedding)
Church of St. Oswald, Arncliffe
St Michael & All Angels Church, Skipton

Field Hospital
Jervaulx Abbey, Ripon

Pandhi Family House
Bradford Industrial Museum, Bradford

Series One–Three Cast and Crew

Abbie Skinner-Biring
Adam Cheetham
Adam Curtis
Adam Hart
Adam Hovell
Adam Johnston
Adam Montgomerie
Adam Rogers
Adam Stokes
Adam Taylor
Adam Trotman
Adrian Rawlins
Aiden Cook
Alain Gales
Alethea Luwero
Alex Howells
Alex Tridimas
Alexandra Harwood
Alexis Platt
Alice Lupton
Alice Northey
Alistair Hopkins
Allison Gruner
Amanda Browne
Amber King
Amy Gibson
Amy Nuttall
Amy Smith
Andreas Nold
Andrew O'Driscoll
Andy Barrett
Andy Ferguson
Andy Gowing
Andy Hay
Andy Merchant

Andy Sellers
Anna Harrison
Anna Madeley
Annabel Bower
Annemarie Lean-Vercoe
Anthea Nelson
Anthony Hennings
Ashley Barron
Austin Haynes
Bea Arnold
Becky Kerton
Becky Summers
Ben Atkinson
Ben Frow
Ben Marshall
Ben Vanstone
Bethan Lennard
Bethany Croft
Beverley Keogh
Billy (Security)
Billy Prentice
Bobby Swerdlow
Borja Berrosteguieta
Brandon Guffog
Brian Clark
Brian Percival
Caius Tabberer
Callum Woodhouse
Caroline Bleakley
Caroline Cooper Charles
Caroline Rowlands
Carrie Clark
Charles Thompson
Charlotte Hillier

Chelsea Roper
Chloe Hardy
Chloë Mi Lin Ewart
Chris Cooke
Chris Kaye
Chris Mann
Chris Marshall
Christopher Barrow
Christopher Hennessey
Ciara McIlvenny
Claire Freeman
Claire Johnson
Clayton Grover
Cleo Sylvestre
Clyde Kellett
Colin Callender
Colin Taylor
Conor Deane
Craig Johnson
Crispin Layfield
Damon Waite
Dan Cain
Dan Dewsnap
Dan Hooton
Dan Leon
Dan Nightingale
Dan Pritchard
Dan Robinson
Dani Biernat
Daniel Shepperson
Daniel Unitas
Danielle Everdell
Danny Janes
Danny Riley
Darren Stuart Neal
Dave Hardy
Dave Hill
Dave Milligan
Dave O'Callaghan

Dave Owen
Dave Reading
Dave Shaw
David Martin
David Mills
David Reading
David Swetman
David Thrasher
Dean Clark
Dean Lee
Debbie Banbury-Morley
Debbie O'Malley
Declan O'Connor
Derek the Pekingese
Derry Wells
Diana Rigg
Diane Hart
Dominic Byles
Dominic Thompson
Donnie Macdonald
Dorothy Atkinson
Drew Cain
Ed Brookes
Ed Glendenning
Ed Miller
Eddy Popplewell
Edith Smith
Edward Borkowski
Edward Dalton
Ella Bernstein
Ella Brewin
Ella Bruccoleri
Elle Crow
Ellie Gillard
Ellie Keighley
Emma Bird
Emma James
Emma Reid
Erik Molberg Hansen

Erik Persson
Estella Harrowing
Euan Macnaughton
Frances Tomelty
Freddy Syborn
Frederic Evard
Fyn Smith
Gabriel Quigley
Gabriela Griffiths
Gareth Proctor
Gareth Williams
Garry Hedges
Gary Barnes
Gary Cadwell
Gary Hoptrough
Gary Redford
Gemma Povey
Gemma Pryke
George Turner
Georgina Vaughan
Gethin Jones
Glenn Lean
Guy List
Guy Slocombe
Hannah Allan
Hannah Reid
Harriet Dale
Harrison Wakeling
Harry Bedford
Harvey Slater-Walker
Hattie Edkins
Heather Horsman
Helen Dickson
Helen Quinney
Helen Sheals
Helen Watson
Helen Williams
Holly Cowan
Hugo Heppell

Ian J. Findlay
Ian Mercer
Ian Rowley
Ian Tully
Imogen Clawson
Isobel Bailey
Isobel Tysoe
Jack Henshaw
Jack Hutchinson
Jackie Sweeney
Jacqueline Smith
Jake Horsfield
Jameel Kazmi
James Baylan
James Burrows
James Clark
James Dean
James Fleet
James Gray
James Pavey
James Pratt
James Swift
James Youd
Jamie Crichton
Jamie Gorza
Jamie Lucas
Jan Sullivan
Jason Bond
Jay Pales
Jeff Brotherton
Jeff Hewitt-Davis
Jenny Farris
Jennifer Johnson
Jenny King
Jeremy Deehan
Jessica Clark
Jessica Gowing
Jill Atkinson
Jill Clark

Jim Moir
Jo White
Joanna Beatty
Joanne Denison
Joanne Tudda
Jody Gordon
Joe Dixon
Joe Martyn
Joe Osborne
John O'Sullivan
John Richardson
John Tueart
John Turner
John Warren
John Wilson
Jolyon Coy
Jon Furlong
Jonathan Eckersley
Jonathan Wyatt
Joseph Allman
Joseph Lee
Joseph May
Jules Jackson
Julian Jones
Julie Edwards
Juliet Rees
June Watson
Justin Hayes
Karen Everson
Karen Wardle
Karim Khan
Kat Pickering
Katie Burnside
Katie Draper
Katie Turner
Katy Metcalfe
Kaye Kent
Kayleigh Platt
Keith Hufton

Kim Vinegrad
Kirsty Binns
Kriss Dosanjh
Kristie O'Brien
Kurt Ryan
Lamin Touray
Lara Steward
Laura Lindsay
Laura Shackleton
Laura Wilkinson
Lauren Taverner
Laurence Good
Lee Martin
Lili Brewin
Lisa Holdsworth
Lisa Parkinson
Lisa Vanoli
Lissa Haines
Liz Armstrong
Liz Lucas
Lizzie Tait
Louise Pedersen
Louise Wilcock
Lucy Crossley
Lucy Smith
Lynda Rooke
Lynsey Palmer
Madeleine Shenai
Maimie McCoy
Marc Pickering
Mark Atkinson
Mark Chatterton
Mark Lisbon
Mark Noble
Mark Rogers
Mark Vaughan
Mark Waters
Martin Day
Martin Radmall

Mary Hanrahan
Mary Wainwright
Mason White
Mat Holloway
Matt Atkinson
Matt Charlton
Matt Cooper
Matt Crook
Matt Hope
Matt Pope
Matt Squire
Matt Turnbull
Matthew Lewis
Max Goldberg
Max Graham
Melanie Kilburn
Melissa Gallant
Metin Hüseyin
Michael Bell
Michael Geary
Michael Maloney
Mick Ward
Mike Green
Mike Harding
Mike Hennessey
Mike Hogan
Mike Richardson
Mita Patel
Mitchell Thomas
Mollie Winnard
Naomi Radcliffe
Natalie Grady
Natalie Lawson
Natalie Mason
Neil Dorward
Neil Hurst
Nicholas Ralph
Nick Gorman
Nick Powell

Nicole Blank
Nigel Betts
Nigel Havers
Noëlette Buckley
Paresh Dayalji
Patricia Hodge
Paul Burnett
Paul Edwards
Paul Hawkins
Paul Hawkyard
Paul Krajewski
Paul Lemming (Splash)
Paul Murphy
Paul Rosato
Paul Testar
Paul Tighe
Paul Walton
Pauline Fowler
Pete Glover
Peter Drinkwater
Peter Roberts
Peter Shaw
Pev Latif
Phil Pease
Philip Hill-Pearson
Phillipa Cole
Phoebe Workman
Pippa Kinsey
Rachel Shenton
Rachel Vipond
Raymond McArthur
Rebecca Eaton
Rebecca Mitchell
Rebecca Stockdale
Rhianne Deans
Rhys Kimmitt
Richard Burrell
Richard Ormrod
Richard Scott

Richard Twigg
Richard Wood
Rob Jarman
Rob Mcgregor
Rob Pavey
Rob Rowley
Robert Marwood
Robin Anson
Rodrigo Griffiths
Rory Ellis
Ros Little
Rosie Lorenz
Ross Douglas
Roxanne Morgan
Russ Youmans
Russell Beeden
Ryan Horsfield
Ryan O'Neill
Ryan Parry
Sam Heron
Samuel Jordan
Samuel West
Sarah Akbar
Sarah Forbes
Sasha Ransome
Scott McIntyre
Seán Carlsen
Sebastian Cardwell
Seumas Mackinnon
Sharon Moran
Shelley Lankovits
Sheraz Ahmed
Simon Rodwell
Simon Wilkinson
Sofia Pamilo
Sonny Sheridan
Sophie Khan Levy
Sophie Mensah
Srdjan Kurpjel

Steffen Goeschel
Stephanie Lalley
Steve Gardner
Steve Jackson
Steve Smith
Steve Yates
Steven Blakeley
Steven Grainger
Steven Hartley
Steven Welsh
Stewart McNicholas
Stewart Renwick
Stewart Svaasand
Susan Cooke
Susan Holmes
Susan Jameson
Susanne Simpson
Tara Llewellyn
Thomas Goodwin
Tim Bain
Tjasa Rahne
Tom Corbett
Tommy Hayes
Tony Jackson
Tony Lucas
Tony Pitts
Tracie Wright
Ursula Haworth
Usmaan Arshad
Vanessa Whyte
Vanessa Zanardo
Vicky Owens
Vivienne Race
Wendy Copeland
Will Lyte
Will Pinnington
Will Thorp
William Oswald
Zach Roberts

Acknowledgements

Alf Wight, the creator of James Herriot, was always an appreciative man. 'He appreciated every little bit of good fortune that came his way,' says Jim Wight, Alf's son. 'He never missed the opportunity to say thank you.' These acknowledgments are a sincere thank you to those who have helped in the making of the series, and to those who have assisted in the creation of this book.

From the producers of All Creatures Great and Small:
Neither the series nor this book would exist were it not for Alf, and so to him we are deeply grateful. Heartfelt thanks are also due to Jim, his sister Rosie Page and their families for affording us the privilege of bringing Alf's stories to the screen once again. We are grateful to Yorkshire – for the landscapes and the hospitality of Yorkshire people. The residents of Grassington have been especially kind.

Sincere thanks to Channel 5 and Masterpiece, All3Media International and Screen Yorkshire. With gratitude to the incredible veterinary profession who take such good care of our animals – we appreciate you. And to the animals – great and small – who were arguably the real stars. A special thank you to James Steen, who has so beautifully and skilfully captured the spirit of the All Creatures experience for you, and whose infectious enthusiasm has made this the most joyous of collaborations. Thank you to our audience – for embracing a new adaptation and giving it such a warm welcome. We made the show for you.

From the author:

This book began, as is the way, with a blank page. Today, I have a laptop brimming to bursting point with tens of thousands of words of notes, draft upon draft of chapters and recordings of interviews with the many specialists, masters of their specific crafts who came together to make the series.

Their voices have filled these pages, and I am indebted to all of them for sharing their stories and insight. You will find them all in this book, and each of them presented a different and fascinating perspective on the making of the series. They are the members of the big 'family' of All Creatures, and I am profoundly privileged to have been the listener.

One can write a manuscript, but it takes many to create a book. I am truly grateful to all of those who were the many. They include:

Nicki Crossley at Michael O'Mara Books. Very kindly, Nicki brought me into the world of All Creatures. Thank you to Louise Dixon and to Shirley Patton for your encouragement. All3Media International made this happen and therefore deserves huge credit, as do cover designer Ana Bjezancevic, typesetter Claire Cater, and proofreaders Ollie Cotton and Gabriella Nemeth. Richard Rosenfeld, the copy-editor and adept at unravelling spaghetti, has tidied up extremely neatly.

Jim Wight has been a guiding light and it has been my good fortune to meet him and hear the tales of his father and the life of vets in Yorkshire. Ian Ashton at The World of James Herriot, in Thirsk, was generous with his time, and one morning gave me an entertaining tour of his wonderful visitor attraction, one of the many highlights of this memorable project. Thank you, Yorkshire lass and Herriot fan Melanie Brown (yes, Mel B), for letting me stay at hers in return for cooking a Herriot-inspired dinner.

The page would have remained blank for quite a while were it not for the enthusiasm, thoughtfulness, constant assistance and collaboration of Melissa Gallant (executive producer) and Sharon Moran (production executive and co-executive producer). Two very special people, I shall always be appreciative of their contribution.

And to you, dear reader – a huge thank you.